Vertical Transportation for Buildings

ELSEVIER ARCHITECTURAL SCIENCE SERIES

HENRY J. COWAN, *Editor*
Department of Architectural Science, University of Sydney

H. J. COWAN, An Historical Outline of Architectural Science, 1966

J. F. VAN STRAATEN, Thermal Performance of Buildings, 1967

DAVID CAMPION, Computers in Architectural Design, 1968

W. FISHER CASSIE, Fundamental Foundations, 1968

H. J. COWAN, J. S. GERO, G. D. DING, and R. W. MUNCEY, Models in Architecture, 1968

J. A. LYNES, Principles of Natural Lighting, 1968

PETER JAY and JOHN HEMSLEY, Electrical Services in Buildings, 1968

B. GIVONI, Man, Climate, and Architecture, 1969

RODNEY R. ADLER, Vertical Transportation for Buildings, 1970

A. BENJAMIN HANDLER, Systems Approach to Architecture, 1970

ANITA LAWRENCE, Architectural Acoustics, 1970

Vertical Transportation for Buildings

RODNEY R. ADLER
Marketing Communications Consultant
Paramus, New Jersey

American Elsevier Publishing Company, Inc.
New York · 1970

AMERICAN ELSEVIER PUBLISHING COMPANY, INC.
52 Vanderbilt Avenue, New York, N.Y. 10017

ELSEVIER PUBLISHING COMPANY, LTD.
Barking, Essex, England

ELSEVIER PUBLISHING COMPANY
335 Jan Van Galenstraat, P.O. Box 211
Amsterdam, The Netherlands

International Standard Book Number 0-444-00072-0
Library of Congress Card Number 73-104976

Printed in the United States of America

Contents

Preface

This work is intended to outline salient vertical transportation principles of significance to the building planner in doing his own job. It aims to clarify his understanding of modern elevator and escalator systems and their possible contribution to building design and operation. A building may then be planned to benefit fully from vertical transportation services and at the same time to accommodate the necessary vertical transport installation.

As with other building services, the costs involved in vertical transportation are crucial to the architect and his client. I have therefore attempted to suggest the economic consequences of alternative plans in terms of relevant cost factors, including some that have so far had scant recognition.

Understanding interrelationships like these—between building and system, service and cost—may not enable the architect to perform the functions of the elevator or escalator expert. But the person responsible for planning the building as a whole will, I hope, gain a firmer foundation on which to evaluate recommendations by specialists.

For nearly two decades the author has been privileged to work with officials of Otis Elevator Company in disseminating information on vertical transportation developments and their application to building requirements.

In the preparation of the present volume, these authorities offered their guidance freely, made available a wealth of illustrations, data and other material, and reviewed chapters on their fields of specialized interest. Without the cooperation so generously provided, "Vertical Transportation for Buildings" could not have been produced.

RODNEY R. ADLER

Paramus, New Jersey
Spring 1970

Chapter 1

Building Design and Vertical Transportation

Around the world, skylines of cities and their suburbs are assuming a new and loftier look. These dramatic changes in the man-made landscape suggest that, of the many systems and services essential to the modern building, vertical transportation may exert predominant influence on its basic design (Fig. 1.1).

The architect's interest in elevators and escalators therefore extends beyond information about equipment to be incorporated in

Fig. 1.1. In and near cities around the world, high-rise buildings made practical by vertical transportation are creating new skylines. This apartment complex is at Meudon-la-Foret, near Paris.

his building plans. At an early stage in their formulation he must now consider potentials of vertical transportation that could significantly affect the very size, shape, and nature of the structure. This volume aims to outline concepts that may guide the architect not only in planning a better vertical transportation system but also in fully using vertical transportation to plan a better building.

1.1. *Influence on Multistory Design*

Since pioneer "safety" elevators first removed age-old limitations on building height more than a century ago, vertical transportation has given the architect increasingly greater freedom in building design.

At first this development meant that the height of a building was no longer limited by the average occupant's ability or willingness to climb stairs. More recent progress in vertical transportation has facilitated the design of multistory buildings to meet increasingly stringent demands for all-around economy and higher standards of service (Figs. 1.2 and 1.3).

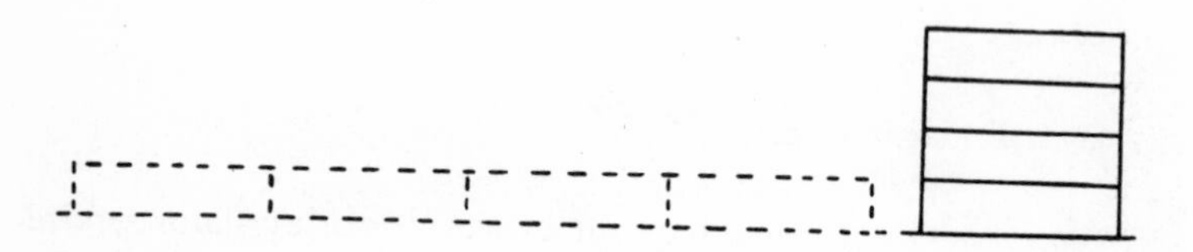

Fig. 1.2. Even where ground space is plentiful, multistory buildings save steps and seconds for their users. High-rise design also transforms more of the space in a building into "outside" space and may free surrounding area for landscaping or other purposes.

Interest in multistory design is now being intensified as land prices rise and desirable sites become harder to find. Costs that sprawling structures exact in horizontal traffic congestion and inaccessible location are receiving renewed recognition. Positive advantages of in-town accessibility for commercial, institutional, and residential buildings are again being emphasized.

High-rise buildings economize ground space and permit more flexible grouping of facilities in a project. Eliminating long, time-consuming walks in low, horizontal structures, effective elevator and escalator service fully realizes the inherent advantages of compact vertical design (Figs. 1.4 and 1.5).

1.2. *Objectives in System Planning*

Improved highways, transit lines, and other forms of horizontal transportation make properties more accessible and therefore more valuable. Within a building, vertical transportation, in contributing to the accessibility of upper-floor space, also enhances its value. Above the first few floors the usefulness of building space depends largely on how well it is served by elevators or escalators.

Adequate vertical transportation for a busy multistory building demands an installation which may account for more than 10 per cent of total construction cost. For maximum return on this investment the system should be planned to provide greatest convenience for the building's users and impose minimum demands on its space and structure.

1.3. *Service Requirements*

Promptness is a prime requisite of elevator or escalator service. Especially in buildings where people work, the value of time-saving transportation increases with rising employment costs.

Fast, frequent service saves time for people waiting for elevators and riding to their floors. By gaining a few seconds for each passenger on every trip, effective elevator or escalator service can, in the course of a year, save thousands of valuable man-hours for all the people in a building.

Space not only commands higher rentals but also is more readily rented in the building with better vertical transportation. Superior standards of service in new or modernized buildings account in part for higher rental and lower vacancy rates than in buildings with outmoded equipment.

Fig. 1.3. Superskyscrapers like the twin 110-story tower buildings of the World Trade Center in the Port of New York become economic with recent advances in vertical transportation. Elevator systems apply the "sky lobby" concept (p. 157) to provide fast, frequent service throughout each tower without taking too much space in its lower portion for hoistways.

1.4. *Planning for Economy*

Rising construction costs and insistent demand for space heighten the value of designing a vertical transportation system to deliver the most service for the equipment installed.

Recent advances in elevator engineering help to attain this goal by saving hoistway and corridor space at each floor and requiring less construction to house machinery. The result is more usable area in proportion to building cube. Elevator and escalator systems can also be designed and located to release usable space where it is more valuable, on upper floors or in outside areas.

Fig. 1.4. Developments to reduce the relative cost of elevator and escalator service encourage its wider application even in lower-rise buildings. Increasing numbers of increasingly active older persons, together with rising standards of service demanded by people of all ages, are creating new demands for vertical transportation.

1.5. *Building Design and Use*

For maximum serviceability and economy, vertical transportation must be planned to meet the requirements of each building and the people who use it.

The size, height, and shape of a building, together with the location of entrances and other facilities, are factors influencing the volume and pattern of vertical traffic. Also significant are the number and distribution of occupants and visitors, and the timing of their arrival, departure, and circulation. Because the complex, variable elements of design and use differ appreciably from one building to another, no two require exactly the same vertical transportation.

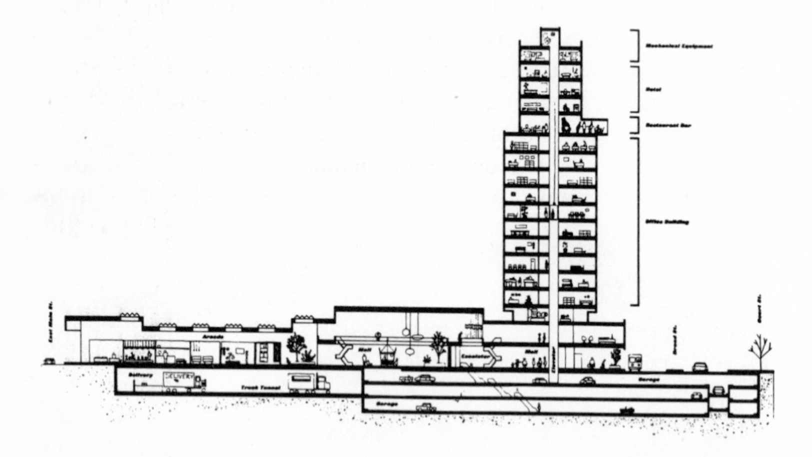

Fig. 1.5. Intown shopping-office-hotel center over a three level garage uses elevator and escalator systems to minimize walking and put its many facilities within minutes of each other.

1.6. *System Design Elements*

A building and its vertical transportation system are reciprocally interrelated; each, to a marked extent, affects the design of the other. Both should be planned concurrently to achieve the necessary integration of building and system.

Considering the building and its vertical transportation system as an entity, the architect can ascertain the total requirements that the system must satisfy. His next task is to determine how the total traffic can best be served, possibly by a combination of elevators and escalators. The performance required of the vertical transportation system is finally translated into the capabilities of individual elevators and escalators.

The type, number, size, and speed of units must be determined, their location and arrangement planned, the methods of operation and control selected. Besides these functional considerations, visual integration of elevators and escalators with the building demands attention.

Although component elements of elevator and escalator systems are standardized, each installation is often individually designed to satisfy the requirements of a particular building. These requirements include not only functional, economic, and esthetic criteria but also compliance with applicable building and safety codes.

Succeeding chapters outline the planning of a building's vertical transportation system and selection and application of its component elements.

Chapter 2

The Demand for Vertical Transportation

Planning vertical transportation for a building begins with determining its needs for service. The *quantity* of service required depends on the volume or intensity of vertical traffic, as measured by the number of people to be carried from floor to floor during a given period of time. Vertical transportation must also satisfy demands for *quality* of service in terms of its promptness and other characteristics.

2.1. *Evaluating Traffic*

Probable patterns of vertical traffic in a new building can be projected with considerable accuracy while the project is still on the drawing board.

For nearly half a century, elevator engineers have been surveying and analyzing traffic in principal types of buildings. From data accumulated for thousands of buildings, anticipated traffic can be related to building population and design and other factors.

Engineers planning vertical transportation systems now use digital computers to process data on all these determinants and create realistic mathematical "models" of possible installations (Section 13.5). Interaction of people and elevators or escalators is simulated under a diversity of conditions, permitting traffic analyses far more comprehensive, accurate, and dependable than when values were calculated manually.

If elevators or escalators are to be installed or modernized in an existing building, vertical traffic can be ascertained by a traffic survey on the premises. However, vertical transportation modernization may be part of a more comprehensive program of building improvement affecting occupancy and traffic. In that case, survey findings may have to be modified by projections of probable service demands in the modernized building.

2.2. *Traffic Volume*

For new or existing buildings, traffic analyses or surveys provide the architect with pertinent information on the volume and distribution of actual or anticipated vertical circulation.

Traffic volume or intensity is a measure of the total number of people demanding transportation between floors every so many minutes. In any particular building, volume may vary over a broad range from moderate to intensive as relatively few or many passengers call for elevator service. When traffic is moderate, elevators, after answering existing calls, may park in readiness to respond to subsequent calls. Intensive traffic is generally characterized by calls in such close succession as to keep elevators in almost continuous operation.

2.3. *Traffic Distribution*

Distribution of traffic as well as its volume should guide the planning of a building's vertical transportation system. Only then can elevators and escalators provide service when and where people want it.

At some times most passengers may be moving between the main and upper floors, while at other times people may be going from one upper floor to another. Traffic may be predominantly up or down, or moving in both directions in varying proportions.

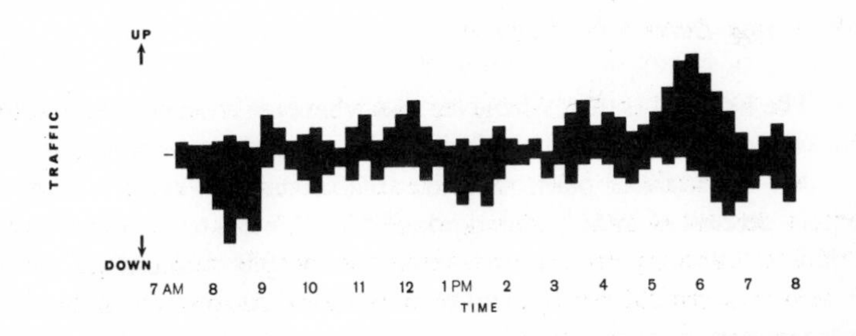

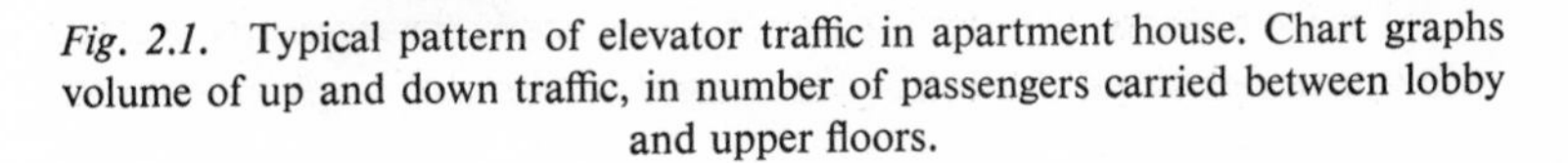

Fig. 2.1. Typical pattern of elevator traffic in apartment house. Chart graphs volume of up and down traffic, in number of passengers carried between lobby and upper floors.

Demand for up and down service varies from moment to moment and from hour to hour throughout the day and night (Figs. 2.1, 2.2, and 2.3). Patterns of elevator and escalator traffic reflect the ebb and flow of people entering or leaving multistory buildings or circulating within them from floor to floor.

In many office buildings and in most apartment houses the heaviest vertical traffic occurs when occupants enter or leave the building. In hospitals, on the other hand, traffic is often heaviest when personnel move from floor to floor within the building.

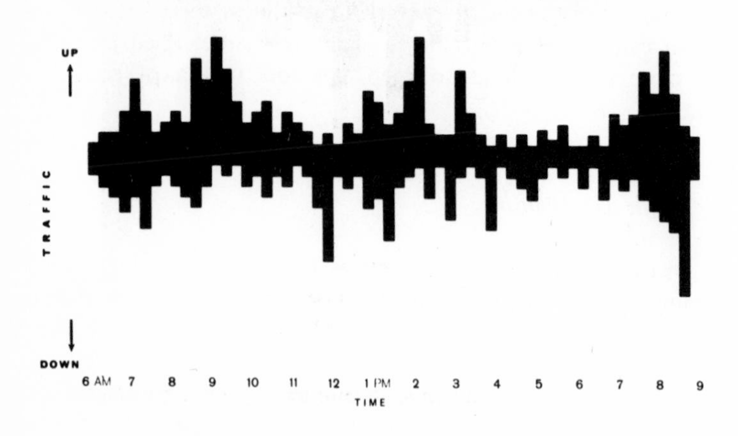

Fig. 2.2. Typical pattern of elevator traffic in general hospital.

2.4. *Traffic Determines Capacity*

The heaviest traffic, whenever and wherever it occurs, determines the capacity required of a building's vertical transportation. Its elevator or escalator plant must be able to carry, every 5 minutes, a certain percent of total building population. Unless the equipment has sufficient capacity to accommodate the people demanding service during this critical period, traffic may cause congestion at landings or on elevators or escalators.

Traffic volume is recorded in 5-minute totals, because this period is long enough to yield meaningful counts yet short enough to measure

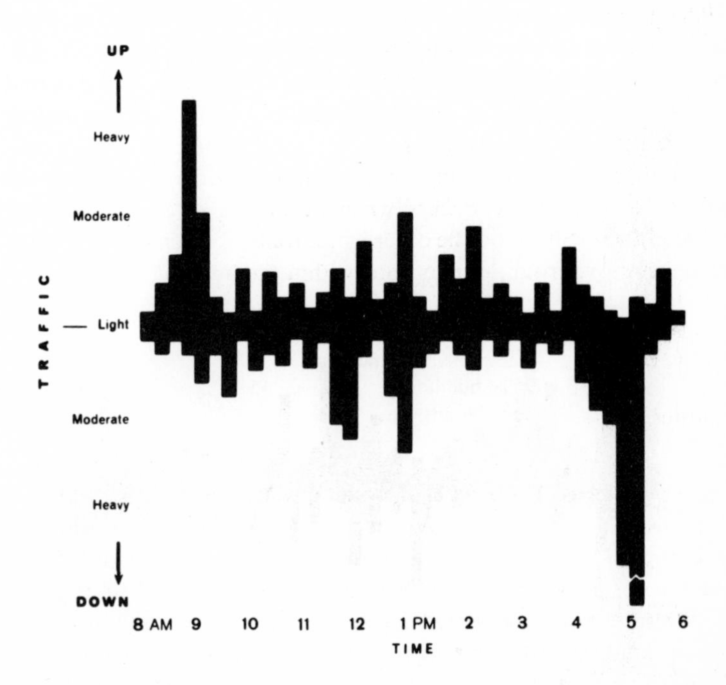

Fig. 2.3. Typical pattern of elevator traffic in office building.

even brief traffic peaks. Elevator handling capacity is also calculated on the basis of 5-minute periods and should at least equal the volume of traffic during the busiest such period.

During this critical 5-minute period, traffic may range from a low of 5 percent of the population of an apartment building to a high of 40 percent in schools. Office buildings have 5-minute critical traffic periods between 10 and 20 percent of their population (Table 2.1).

2.5. *Factors Determining Traffic*

Since a building of specialized type is used by its occupants in a correspondingly specialized way, their movement into, within, and out of buildings or a particular type of building tends to assume dis-

Table 2.1
Five-minute peak traffic as percent of building population above first floor

Apartment building	5–7%
College residence	10–15%
Department store (customers and staff)	15–25%
Hospital, general (all pedestrian traffic)	10–15%
Hotel or motel, in-town	7–12%
Office building, diversified tenancy	12–15%
Office building, single-purpose	15–25%

tinctive patterns. Each type of building—commercial, institutional, or residential—therefore has characteristic patterns of vertical traffic, as Figs. 2.1, 2.2, and 2.3 suggest. Traffic peaks and lulls differ from one type of building to another in intensity, shape, and timing.

Despite the general similarity of patterns for buildings of a certain kind, no two buildings, even of the same type, have exactly the same vertical traffic. Its volume and distribution are influenced by the height, size, and design of each individual building.

Sheer-rise towers (Section 10.5), for instance, have greater net areas on upper floors. Traffic to those floors is usually heavier than in buildings of setback design.

Special traffic-generating facilities, from basement garages to rooftop restaurants, may need extra elevator or escalator service (Figs. 2.4 and 2.5). A building with large tenants, each occupying several floors, often experiences considerable interfloor traffic. The trend to completely controlled indoor environment may result in more intensive use of space on certain floors and, consequently, heavier vertical traffic.

Conditions outside the building also influence the pattern of vertical traffic within its walls. A major external factor is the community's horizontal transportation—private automobile or mass transit—that delivers people to the building. The rate at which they arrive and depart helps to determine the elevator or escalator service needed inside. Retail shops and other attractions, near but not necessarily in a building, may also influence its vertical traffic.

As internal and external conditions change over the years, a building's traffic pattern changes accordingly. Vertical transportation in the building must therefore be capable of adaptation to changing demands.

Fig. 2.4. In Kuala Lumpur, Malaysia, the 13-story Chartered Bank has drive-in banking on the ground-floor level, with two escalators to the main banking floor above. The building also has four passenger elevators, a service car, and a bullion lift. Booty, Edwards & Partners, architect; Steen Sehested & Partners, consulting engineer; Frank & Vargeston, quantity surveyor; Yew Lee & Company Limited, contractor.

Fig. 2.5. Frankfort Intercontinental Hotel, Frankfort-on-Main, has 500-car basement garage and 21st-floor restaurant and roof garden. System of four passenger and two service elevators provides transportation to these as well as to guest-room floors and other facilities. Apel & Becket, and Becker Ingenieur, Frankfort, architects: Messrs Philipp Holzmann A.G., Julius Berger A.G., and Siemens-Bauunion G.m.b.H., general contractors.

2.6. *The Demand for Service Quality*

Elevator transportation may be sufficient in quantity to handle all passengers who demand service during every 5-minute period, yet may keep some of them waiting too long. Service quality depends partly on the frequency of operation, which determines passenger waiting time.

Excessively long waiting and riding time is not only annoying to the individual passenger but also costly to his employer. At today's high and still-rising levels of salary and fringe costs, organizations that use buildings—owned or rented—demand fast, frequent vertical transportation that conserves employee and executive time.

Service frequency is measured by average time intervals between elevators in either direction. Short intervals reduce the time lost by passengers in waiting for and riding on elevators.

While acceptable service intervals vary in general with the type of building (Table 2.2), specific circumstances influence intervals acceptable to the people using a specific building. If it commands luxury-level rentals, for example, tenants tend to be more critical of service quality. Expected standards are higher in some cities than in others.

Table 2.2
Desirable service frequencies, in average interval of time between elevators in the same direction

Apartment building	50–70 sec
College residence	50–70 sec
Department store	30–50 sec
Hospital, general	30–50 sec
Hotel or motel, in-town	30–50 sec
Office building	20–35 sec

Passengers prefer service at intervals that are not only short but also consistent; excessive, irregular waits bring complaints. To conserve riding as well as waiting time, service must be fast as well as frequent. It must also be as direct as possible, since passengers dislike too many intermediate stops.

With the aid of traffic surveys or analyses, the architect and engineer, in consultation with the owner, can ascertain the quantity and quality of vertical transportation a building requires. Whether the necessary service can be provided most economically by elevators, escalators, or a system combining both forms of transportation calls for evaluation of their relative merits as outlined in the following chapter.

Chapter 3

Satisfying Service Requirements

Having analyzed actual or anticipated requirements for vertical transportation in a building, the architect may begin planning an elevator or escalator system to provide service of the required quantity and quality. Pertinent data on which to base system design, as Chapter 2 has indicated, include the amount of vertical traffic and its distribution and location.

To handle the traffic, the architect may visualize one or more rising and descending platforms, their number, area, and speed adequate to carry the passengers wishing to ride in a period of time. A larger number of smaller platforms is generally preferred to satisfy requirements for service quality as well as quantity (Sections 2.4 and 2.6). If the platforms are so numerous that they arrive continuously, as in the escalator, each passenger enjoys immediate service.

Ascertained requirements are met by selecting elevators or escalators of appropriate specifications and integrating them into a system with the desired performance. Understanding the capabilities and limitations of these two basic types of vertical transportation equipment is essential to the competent planning of an installation.

3.1. *Elevators or Escalators*

In planning a particular building, "elevators, escalators, or both" can be decided only by evaluating specific conditions in the light of equipment performance, cost, and other characteristics. For trips to floors above the eighth or ninth (or rises greater than 80 or 90 ft), the factor of travel time strongly favors elevators.

If traffic is infrequent, economic factors also point to elevators, even in low-rise buildings. Elevators are often less expensive, not only in equipment cost but also in space consumption. With completely automatic elevators handling even the heaviest traffic, operating costs for elevators and escalators are comparable.

3.2. *Escalator Application*

Escalators provide continuous service and can handle traffic in steady streams for limited vertical distances.

Short-rise escalators may join areas on intermediate levels. A motor inn at Orlando, Florida, has one of the shortest escalators, with a vertical rise of only 5 ft. $6\frac{3}{4}$ in., to take guests to a raised level of the lobby.

Versatility for traffic that varies sharply from moment to moment or floor to floor is another virtue of escalators. They can be reversed so that a pair will provide two-way service when required or both will operate in one direction for peak traffic.

In store buildings, shopping centers, and other applications calling for visibility as well as mobility, escalators have long held the spotlight. But they now also serve many other types of buildings for part or all of their vertical transportation.

3.3. *Elevator-Escalator Systems*

In escalatored buildings with heavy interfloor traffic over several stories, excessive travel time may cause considerable cumulative loss of employee man-hours. On the other hand, even the best elevator system has a measurable average waiting interval, whereas escalator service is continuous.

Both elevators and escalators may therefore be integrated into a combined vertical transportation system in accordance with the characteristics of both forms of service and the total traffic load. Such a system may depend on escalators primarily to serve the lower floors of a building, on elevators for the upper floors (Fig. 3.1).

To fill or empty such a building rapidly during up or down peaks, all escalators would operate in the same direction while elevators would run express to upper floors. At other periods, escalators would operate two-way and elevators would serve all floors, for time-saving interfloor travel.

Investment economy may favor escalators for certain floors. If two lower elevator terminals, like a basement garage and a main-floor entrance, generate appreciable traffic, escalators or separate elevators between those floors may cost less than increased capacity for the main elevators. Where escalators can handle the bulk of passenger traffic, a service elevator is often added for freight and handicapped persons.

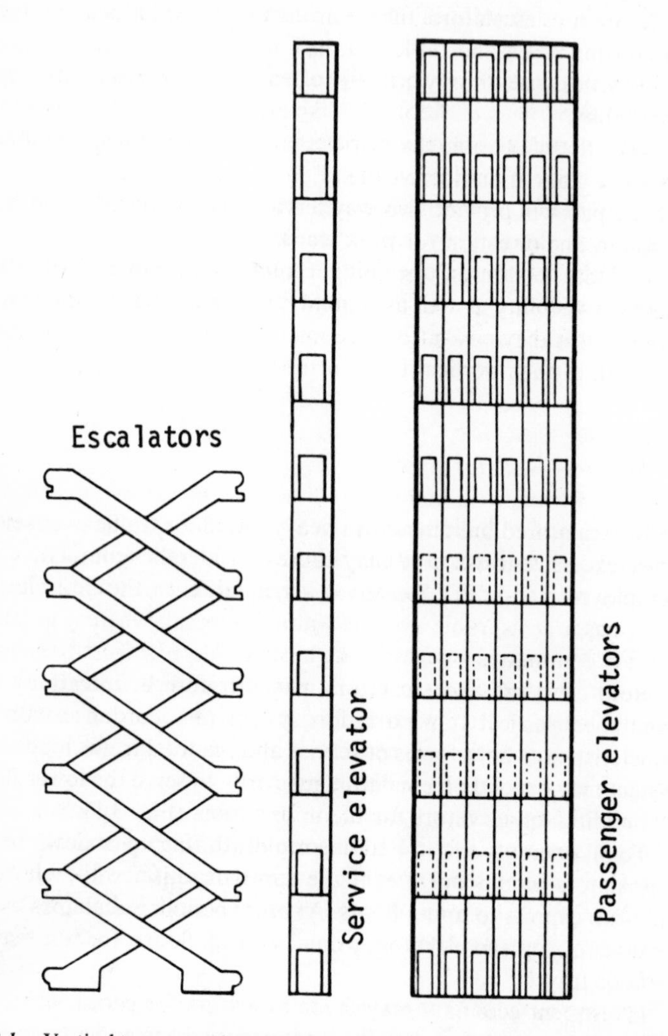

Fig. 3.1. Vertical transportation system combining elevators, primarily for upper floors of a building, with escalators for lower floors.

3.4. *System Location*

In relation to the building plan, elevators and escalators should be located for the most convenient service to the greatest number of people. With other utilities the vertical transportation system usually forms a central service core.

Escalators, like elevators, are often placed centrally for ready access to all parts of all floors. Unlike elevators, escalators need not be stacked vertically but may be placed where they best fit into traffic lanes from level to level.

Occasionally, for open floor areas that may be divided in various arrangements dictated by tenant needs, elevators, escalators, and other utilities may be on one side, in a separate service tower. Planning the installation is also influenced by whether the building is of sheer-rise or setback design.

3.5. *Grouping Elevators*

For frequency and continuity of service, elevators are installed in groups rather than separately. With a single elevator, operating interval equals round-trip time; this results in excessive delays for passengers who just miss the elevator and must wait for it to come around again.

When two or more elevators are located side by side and operated as a group, average interval equals round-trip time divided by the number of cars; shorter waiting time results. Another advantage of grouped elevators is that, if one car is shut down for maintenance, passengers are not entirely without service.

If a building is taller than 15 or 20 stories and requires a large number of elevators, vertical transportation efficiency and economy may be improved by separating the cars into express and local groups (Fig. 3.2). With each group serving only some of the building's floors, elevators tend to make fewer stops and express cars can attain higher speeds.

Separate elevators for service traffic are proving their worth in commercial (Section 6.7) and institutional buildings (Section 8.2). Busy office buildings benefit from service elevators with wide, power-

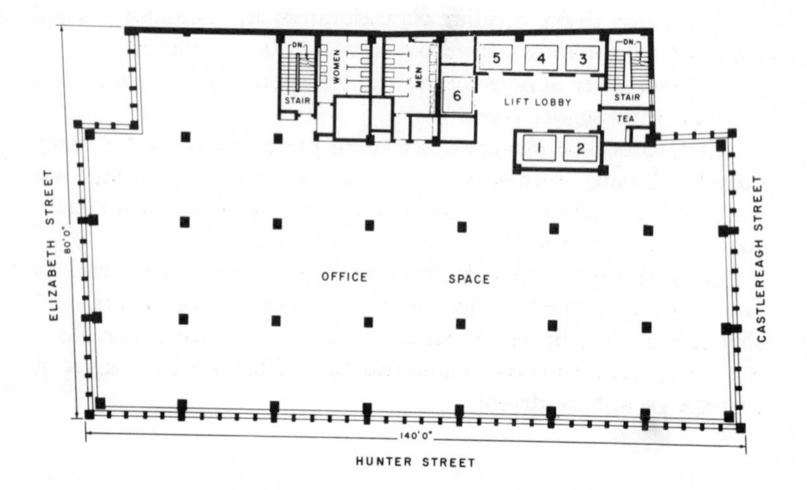

Fig. 3.2. P & O Building, Sydney, has five 3,000-lb. passenger elevators and a 3,500-lb. service car (6) which can also carry passengers during busy periods. Fowell, Mansfield & Maclurcan, architect; T. C. Whittle Pty. Limited, builder; John R. Wallis, Spratt & Associates, mechanical engineer; Woolacott, Hale, Bond & Corlett, structural engineer.

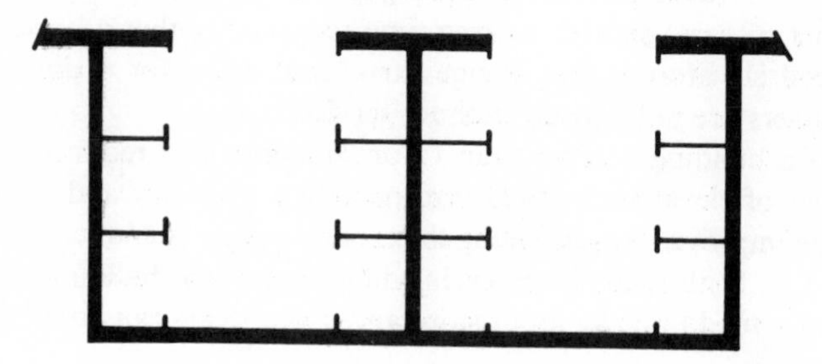

Fig. 3.3. Alcove arrangement for elevator groups, one for local and the other for express service.

operated doors that can handle even furniture and partitions, avoiding interference with passenger traffic. Service elevators may be arranged to add their capacity to that of the passenger cars during traffic peaks (Fig. 3.3).

Measures like these reduce elevator round-trip time, improving transportation in quantity as well as quality. Elevator groups can be planned to provide more or less service to lower or upper parts of the building, as required by its design or occupancy.

3.6. *Elevator Layout*

Most passenger service applications require elevator cars wider than they are deep, with wide entrances and center-opening doors to let people in and out faster. In planning the elevator core of a building, an architect arranges corridors and hoistways in a layout designed to reduce distances waiting passengers must walk to an arriving car.

In a group of two to four elevators, cars may be arranged in a line or facing each other. Five or more elevators should not be located in a single line because the time required for passengers to walk from the ends of the group causes them to miss elevators or delays service.

Elevators in groups of four to six cars are usually laid out in "alcove" arrangements (Fig. 3.2). This plan not only lets passengers wait near elevators but also prevents interference between waiting passengers and people passing through the main corridor.

Groups of seven or eight cars are arranged, not in a closed-end alcove, but preferably on both sides of an open-ended elevator corridor. Alcove or corridor arrangements are common for the elevator cores of larger buildings, with more than one group of cars (Fig. 3.2).

3.7. *Escalator Arrangement*

So that riders can easily transfer from one unit to the next, escalators are installed in crisscross or parallel arrangements (Fig. 3.4). The first plan is more economical of space, the latter is more impressive in appearance.

In either arrangement, escalators may be adjacent or separate.

Newels should be placed far enough apart that passengers leaving one escalator do not interfere with those entering the adjacent unit, and persons going several floors need not make an abrupt about-face each time they change from one escalator to the next.

Express escalators made their appearance in the William A. Shea Stadium in New York. The structure has twenty-one escalators arranged in five express and three local banks, the express escalators skipping two lower levels served by local units.

Express escalators hold interesting possibilities but are unlikely to extend appreciably the total rise for which escalator service is practical. Passengers spend most of their top-to-bottom travel time in actually riding on escalators rather than transferring from unit to unit. Express systems would save only transfer time.

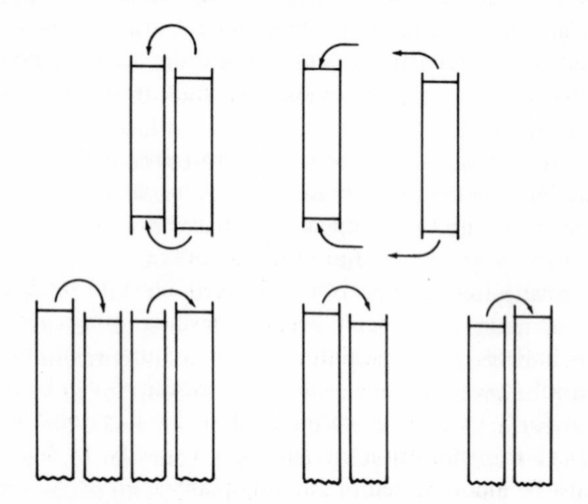

Fig. 3.4. Escalators may be installed in criss-cross or parallel arrangements, with up and down escalators adjacent or separate.

3.8. *Carrying Capacity*

Elevators or escalators of the proper number, size, shape, and speed are selected and integrated into a system with the necessary carrying capacity.

The number of people a system can serve during a given time period, as every 5 minutes or every hour, depends on the number, size, and speed of the moving platforms in the system. If the platforms are elevators, their effective carrying capacity is increased if they fill with people at one floor and take them all directly to another floor, and if passengers get in and out of the cars with the least delay.

The volume of traffic a group of elevators can handle during a 5-minute period depends on the average number of passengers each car carries on each trip, multiplied by the number of round trips in 5 minutes and by the number of cars in the group. Instead of a few very large elevators which would take too long to load and unload, a greater number of smaller cars provides the necessary carrying capacity with more frequent service.

Higher speeds increase effective capacity, because faster elevators can complete more round trips in a period of time. In practice, speeds are limited by the rate of acceleration and deceleration people find comfortable and by the length of nonstop runs on which elevators can attain top speeds.

Escalators can usually carry 2,000 to 6,000 persons per hour, depending on step area and speed. An elevator can serve as many as 200 people in 5 minutes, depending on the complexity of traffic, number of stops, total rise, and other factors.

But generalized comparisons between elevator and escalator handling capacities are difficult, because elevator ratings are based on people in batches while escalator ratings assume continuous flow. Traffic-handling capacity of elevators and escalators can be compared realistically only by an engineering analysis for a specific building.

Vertical transportation of sufficient capacity to handle peak traffic must be incorporated in building plans at an early stage. Since elevators or escalators can be added or enlarged only at considerable expense once a building has been completed, the importance of prior traffic analysis and system planning is apparent.

Chapter 4

The Installation in the Building

After evaluating the principal forms of vertical transportation for their impact on building design and use, the architect may wish to consider the major elements of an elevator, escalator, or electric dumbwaiter installation. His primary concern is not with details of equipment construction and operation but rather with requirements for building space, structural support, electric power, and other provisions necessary to accommodate the transport system.

Architectural plans allocate space for elevator and dumbwaiter machine rooms, hoistways and pits, and wellways for escalators. The building structure must support static, dynamic and impact loads; its electrical distribution system must supply energy to control as well as to operate machinery.

Requirements depend on the type of installation and the specific ratings of its components. Necessary specifications for each piece of equipment are available from manufacturers.

4.1. *Traction Elevators*

Modern elevators for higher rises and speeds are usually of the traction type, whereas hydraulic elevators are favored for low rises and heavy capacities.

In traction elevators (Figs. 4.1 and 4.2) the car is suspended from one end of a set of steel cables, called "ropes," with a counterweight at the other end. The counterweight is about as heavy as the car with a 40 percent load. In the machine room, usually atop the hoistway, the ropes are hung on a motor-driven sheave, with a groove for each rope.

When the sheave turns, traction moves the ropes, causing car and counterweight to travel in opposite directions. Ropes are of such length that, should the car continue rising past the top landing, the counterweight would touch bottom and traction between sheave and ropes would be lost. The traction elevator is thus inherently safe from the possible hazard of being pulled into the structure over the hoistway.

4.2. *Gearless Elevators*

High-speed traction elevators, for speeds from 350 fpm to 1,800 fpm and higher, have driving machines of the gearless type (Fig. 4.1). This name is derived from the fact that traction sheave and drive motor armature are mounted directly on one solid forged steel shaft. A gearless traction machine has three main components: motor, sheave, and brake.

Delivering power to the hoist ropes without gears, the gearless machine is highly efficient. Well-made machines run quietly and last almost indefinitely because they have few wearing parts and their motors rotate at relatively slow speeds. To attain the smoothness of starting, stopping, speedup, and slowdown of which gearless traction machines are capable, they are usually controlled by variable voltage systems (Section 5.2).

In buildings where elevator service of the highest quality is desired, the tendency is to install gearless machines of greater power than formerly to further reduce elapsed time from one stop to the next. More powerful machines more rapidly overcome the inertia of car and counterweight and accelerate the elevator to full speed in the shortest time consistent with passenger comfort.

Above each elevator hoistway the machine room or penthouse must accommodate driving machinery and control equipment. Room should also be left for servicing, removing, and replacing equipment. Ventilation, possibly with air-conditioning system spill air, should be sufficient to remove heat dissipated by elevator machines and controllers and keep machine room temperature between 50° and 90°F.

Typical of higher-speed elevators, the gearless installation shown has hoisting ropes wrapped around not only the drive sheave but also a secondary sheave for increased traction. The secondary sheave is often installed on a level below the machine room floor. This arrangement is called double wrap roping, in contrast with single wrap roping, commonly used for slower, geared elevators (Fig. 4.2).

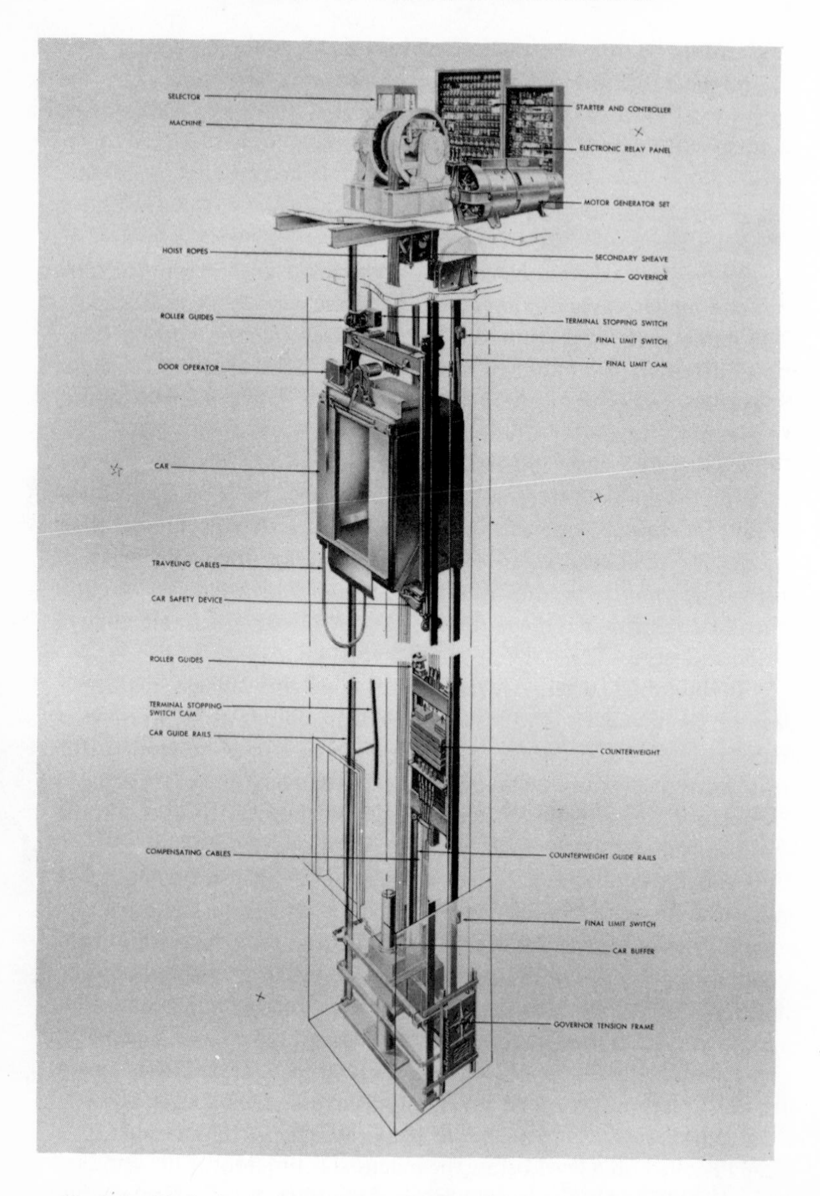

Fig. 4.1. Typical passenger elevator installation with gearless traction machine. Note secondary sheave for double-wrap roping.

4.3. *Car and Counterweight in Hoistway*

The elevator car or cab (Section 11.1), together with the platform, door-operating mechanism, and safety equipment are mounted on the car frame. Structural and operating elements are principally of steel, but rubber blocks between car frame and platform provide sound and vibration isolation for riding comfort.

Efficiency, safety, and smoothness of elevator operation have been improved by the transition from sliding to roller guides mounted on the car frame and running on the guide rails in the hoistway. Rolling motion reduces friction and eliminates the need for guide-rail lubrication, which had been a fire hazard. Roller guides have individual suspension to cushion jolts from uneven rail alignment, and rubber tires to minimize noise and vibration.

In many traction elevators the hoisting ropes, of high-grade traction steel fabricated especially for elevator service, are attached at one end to the top of the car frame and at the opposite end to the steel frame of the counterweight. In other arrangements the ropes run over sheaves on the car and counterweight as well as the driving sheave and are dead-ended at the top of the hoistway.

With car and counterweight near the top or bottom of the hoistway in a higher-rise elevator, the weight of the ropes is sufficient to effect car-counterweight balance. To maintain this balance, compensating cables or chains are attached to the bottom of the car and counterweight and passed under a compensating sheave in the elevator pit. Flexible traveling cables, from the car to a point usually midway up the hoistway, contain electrical conductors for car control, lighting and ventilation power circuits.

4.4. *Safety Provisions*

From the beginning, elevator development emphasized safety. Modern elevators and escalators are designed, electrically and mechanically to prevent situations of possible hazard to passengers. Vertical transportation today is safer than any other means of travel.

An integral part of every traction elevator is the safety system (Fig. 4.3) which prevents the car from overspeeding downward. This

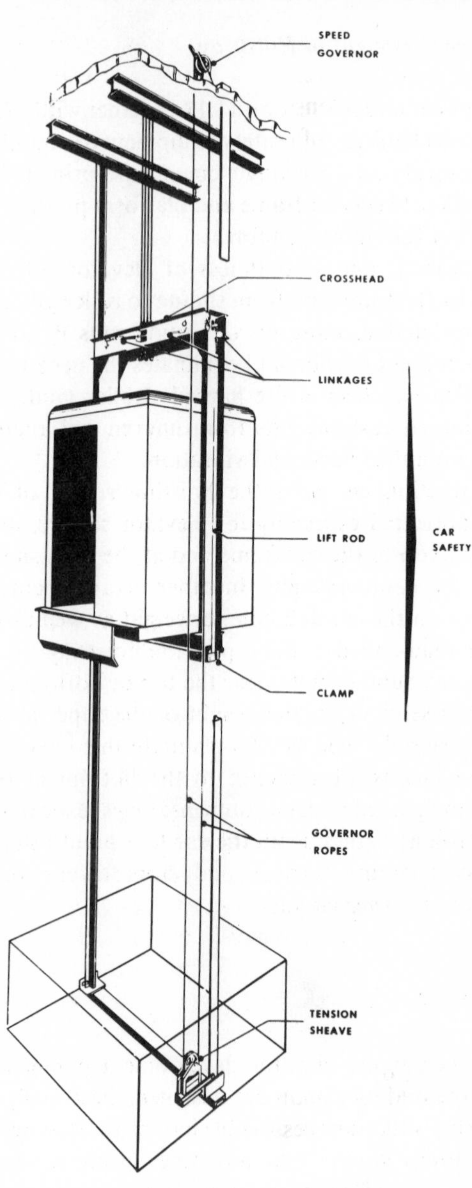

Fig. 4.2. Safety system to prevent elevator overspeeding in down direction.

system, entirely separate from the driving machinery and service brake, consists of a speed governor, a "safety" attached to the car, a steel governor rope, and a tension sheave. The governor rope, both its ends attached to the car safety lift rods, passes over the driving wheel of the governor at the top of the hoistway and under a tension sheave at the bottom of the hoistway.

As long as the elevator speed does not exceed the governor tripping speed, the safety system does not affect elevator operation. Should the elevator begin to overspeed, however, the governor opens a safety switch and cuts off power to the hoisting machine. This initial action applies the service brake and usually stops the car.

If the car's downward speed should continue to increase, the governor trips and clutches the governor rope with enough force to set both safety clamps on the car, forcing the wedge-shaped safety jaws to grip the guide rails. Further motion of the car causes these movable jaws to wedge themselves between the rail and the arms of the clamps until the heavy safety springs exert sufficient clamping to bring the car to a smooth, sliding stop.

After the car stops, the safety jaws continue to grip the rails and hold the car stationary. When the cause of the overspeed has been corrected, the safety may be released by running the car up a few inches, reversing the wedging action, and returning the safety jaws to their "ready" position.

Protection against overtravel is provided by limit switches near the upper and lower ends of the hoistway. A car continuing to move past the top or bottom landing trips a limit switch, cutting off power to the driving machine and applying the service brakes. Should an elevator continue to descend past its lowest landing, the car is brought to a stop by the spring-return oil buffer in the pit. Oil buffers are built to absorb the impact of a loaded elevator descending at full rated speed. A similar buffer is installed under the counterweight.

Elements of automatic control systems like self-leveling (Section 5.3) and electronic door detectors (Section 5.4) guard against accidents at elevator entrances. Modern automatic elevators are even safer than their manually operated predecessors.

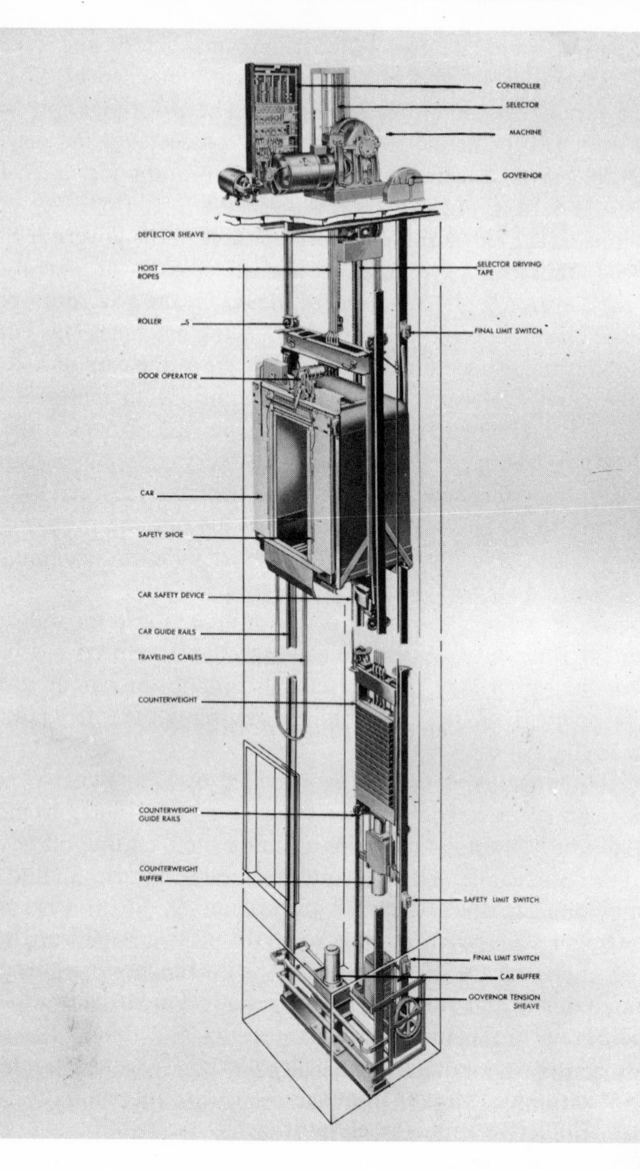

Fig. 4.3. Typical passenger elevator installation with geared machine.

4.5. *Geared Machines*

For speeds under 350 fpm, traction elevators are driven by geared machines, in which a worm and gear transmits power from the driving motor to the drive sheave. A geared-machine installation, with single-wrap roping (Fig. 4.2), incorporates equipment of the same basic type as the higher-speed gearless elevators and operates in an essentially similar way.

4.6. *Special Designs and Materials*

To meet unusual service requirements, elevator equipment may be built of special design or materials. Drive machinery and control equipment, for example, may be capable of operating in explosive atmospheres. Cars and hoistway installations may be designed to operate in exposed locations.

In a building with sensitive apparatus or instruments which could be affected by the magnetic properties of steel, an elevator of nonmagnetic metals may be installed. In one research laboratory the elevator car, car frame, and counterweight frame are of aluminum alloy. Counterweights, usually cast iron, are lead, while hoist and governor ropes have hemp cores wrapped with layers of nonmagnetic austenitic stainless steel.

4.7. *Hydraulic Elevators*

Minimum demands on building space and structure are notable advantages of the modern hydraulic elevator (Fig. 4.4), electrically powered and automatically controlled. Because practical travel speeds are limited to 200 fpm or less, hydraulic elevators are usually installed in buildings up to six stories high.

A steel piston moving in a vertical cylinder as deep as the rise of the elevator supports the car from below. An electric pump forces oil into the cylinder to raise the elevator, while electrically controlled valves release oil into a reservoir to allow the car to descend.

Hydraulic elevators require neither penthouse nor overhead support. The machine room can be located in the basement or other

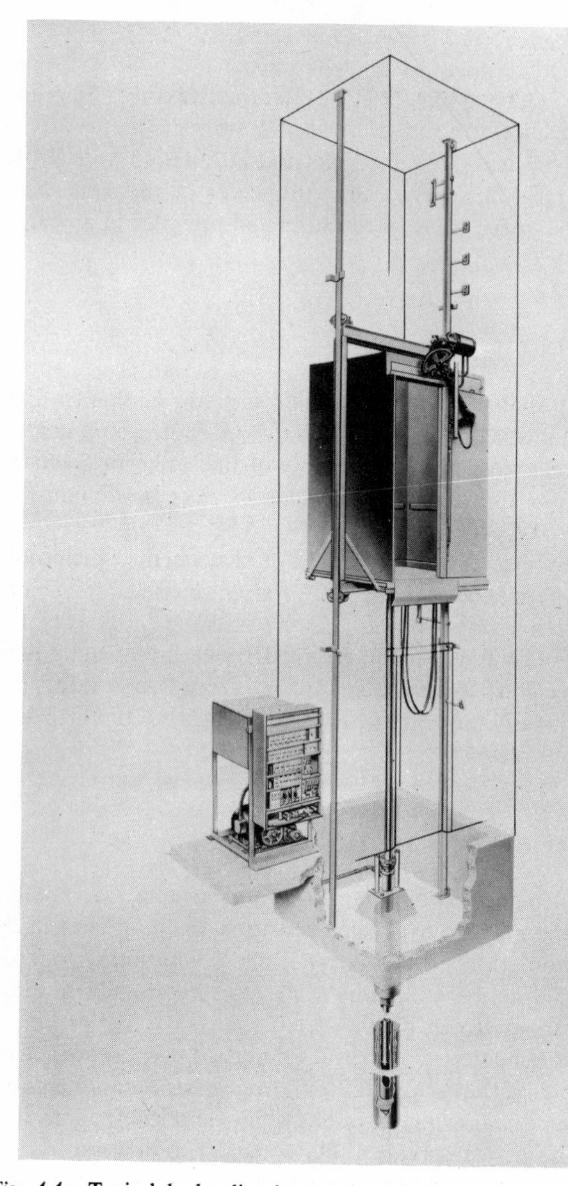

Fig. 4.4. Typical hydraulic elevator installation. Plunger supports car from below, eliminating need for overhead machine room.

lower level, on any side of the hoistway or at a distance from it. For heavy-duty freight service, hydraulic elevators have been installed with capacities of 100,000 lb. or greater.

4.8. *Controlling Elevator Noise*

Although elevators, like other electromechanical equipment in buildings, have long made noise, recent trends in building construction, design, and utilization have called increased attention to the problem and its control.

Higher construction costs have encouraged use of new techniques whereby buildings with generally excellent characteristics can be completed at economically acceptable prices. These methods often produce a building structure more likely to transmit rather than absorb noise and vibration, including sound energy from the elevator installation.

At the same time, residential as well as commercial buildings are going higher (Section 7.4). Greater height necessitates faster elevators, and higher elevator speeds tend to raise noise levels. Elevator sounds that might not be objectionable in a 60-story office building could bring loud complaints in an apartment tower of the same height, where people live and sleep.

In setback buildings, machine rooms for elevators serving lower floors of a building were often on the terraces formed by this design. But in the newer sheer-rise towers (Section 10.5), all machine rooms except those for the highest-rise elevators are usually surrounded by occupied space.

Completely air-conditioned buildings now prevail, with windows either permanently sealed against outside noise, or at least closed much of the year. Shutting out street noise makes occupants more sensitive to noise from inside sources, including the elevators.

Equipment operating in the machine room and cars moving in the hoistway are principal sources of noise in a traction elevator installation. In the slower hydraulic elevator, noise may come primarily from the motor-driven pump and the hydraulic cylinder. Noise may be transmitted to occupied spaces—and the ears and nerves of occupants—through the building structure or through the air.

4.9. *Separating People from Noise Sources*

Space relationship in a building may be planned to minimize the impact of noise on people. Recent commercial buildings with elevators in a separate service tower, primarily to leave large, clear floor areas, enjoy effective separation of occupied areas from possible elevator noise. In the more usual arrangement, a utility core centrally located within the main structure, elevator groups should be surrounded by stairways, service closets, and other installations that also help to absorb sound.

Occupied areas may be so planned that those where noise is most critical are farthest from the elevators. In an office building a mechanized accounting group, for example, could adjoin elevator hoistways, with executive offices in a naturally quieter zone. Apartment or hotel bedrooms may be located away from hoistways or machine rooms.

But spatial separation is not always possible. Tenants of premium-rental penthouse apartment or office suites, for instance, are not likely to tolerate noise from nearby elevator machinery. Sound-transmission paths must then be located and effectively interrupted.

4.10. *Controlling Machine-Room Noise*

In a "captive" elevator machine room, one surrounded by occupied space, the building structure may transmit machinery noise. Support beams for apartment or office floors tied to building beams to which the elevator machine beams are attached form a transmission path.

Sound isolation of gearless elevator machines is more economically planned as part of the original installation, since field labor costs may become appreciable if the work is attempted after machines are in service. A machine is set on rubber pads inserted between its bedplate and the machine support beams of wide-flange design. Setting the beams in correct position, horizontally aligned with each other, evenly distributes the load of the machine over the rubber pads.

Geared machines, motor-generator sets, and controllers may be similarly isolated from the building structure. But mounting a geared machine to minimize transmission of vibration to the surrounding structure does not eliminate airborne noises from worm and gear misalignment. The worm and gear should be realigned if possible, and replaced if badly worn.

Air transmission of noises from equipment in a captive elevator machine room may prove objectionable to occupants of surrounding spaces. Possible sources of sound include rotating elements of driving machines or motor-generator sets, brushes, switches, and brakes. Whatever the source, the solution often lies in acoustic treatment of the machine room.

Airborne noise is best absorbed by a double wall of soft "chalk" tile on the sides of the machine room next to occupied space, with dead air space between the two walls. Doors through the double wall should also be double, installed in tandem, with dead air between them.

A more economical but less effective means is to cover ceiling, wall, and door areas with panels of rock wool or similar soundproofing material, with aluminum facing or other retaining material covering the machine-room side. Acoustic material must be fireproof when installed and remain fireproof throughout its service life.

Air-conditioning ducts may transmit noise from captive machine rooms to other parts of the building. In such cases a single-ply canvas section inserted in the metal duct where it enters the machine room has been found to reduce noise transmission through the duct system.

In some traction elevator installations, the secondary sheave is located in a space just below the main machine room (Fig. 4.5). A hard tile wall around this space may create a sound box which amplifies noises from the ropes running on the secondary sheave. Much of the noise is transmitted down the hoistway, the easiest escape path.

One way to control noise from this source is to eliminate part of the walls (Fig. 4.5, broken lines) under the building support beams. This change opens a larger space in which the sheave noises are dissipated before they can reach the highest occupied floor of the building. If the secondary sheave space cannot be opened up, the inside of the walls on all sides should be covered with sound-absorbing material.

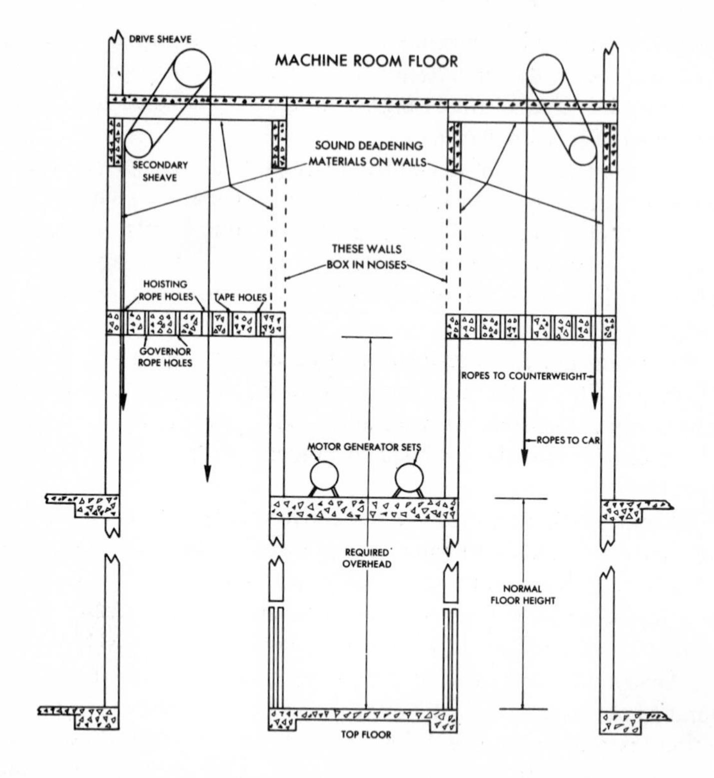

Fig. 4.5. "Sound box effect" may direct secondary sheave noise down hoistway. Noise can be harmlessly dissipated by opening up walls around sheave or by lining space with sound-absorbing material.

4.11. *Controlling Hoistway Noise*

Machine-room noise is likely to bother people only on the same floor, and possibly on the floors directly above and below the room. But noise generated in hoistways and carried by the building structure or through the air can cause complaints on every floor an elevator serves.

In one high-rise building of slip-poured concrete construction with integral walls and floors, the structure was found to be transmitting noise from the movement of the elevator counterweight on its guide rails. Inserting rubber blocks between the rails and rail brackets provided effective sound isolation.

A high-speed elevator traveling in its hoistway can produce air noise. Fortunately, the noise tends to be noticeable only in a single-elevator hoistway and not in multicar adjoining hoistways open to each other, a more usual arrangement in tall buildings served by grouped elevators.

An elevator moving rapidly in a single hoistway builds up air pressure ahead of the car, forcing air to flow around the car and creating turbulence and noise. Varying directly with elevator speed and height of rise, the noise seldom becomes appreciable in buildings under 30 stories.

To reduce noise in one high-rise building with a single-car hoistway, part of the concrete wall in the elevator pit was removed to let air under the down-traveling car escape to adjacent hoistways. Relief was obtained in another building by removing the lower portion of the wall between two adjacent single hoistways so that air could flow from one to the other.

With the more usual double or triple hoistway, air noise is seldom troublesome at present elevator speeds. But, when they become much higher, cars passing each other in a hoistway might cause brief but loud bursts of sound.

Elevators traveling at projected speeds of 2,400 fpm, for example, would rush past each other at nearly 60 mph and might, in an enclosed hoistway, create unpleasant sound effects. They could be reduced by aerodynamic aprons of fiber glass or metal on the bottom and top of the car, like those already in use on the Space Needle in Seattle and other outdoor installations.

Because sound control is still largely an empirical art, situations requiring special treatment can often but not always be anticipated in the planning stage. Design of an installation with the desired acoustic, architectural, and functional characteristics is facilitated by early consultation of the architect with the elevator engineer and the acoustical engineer.

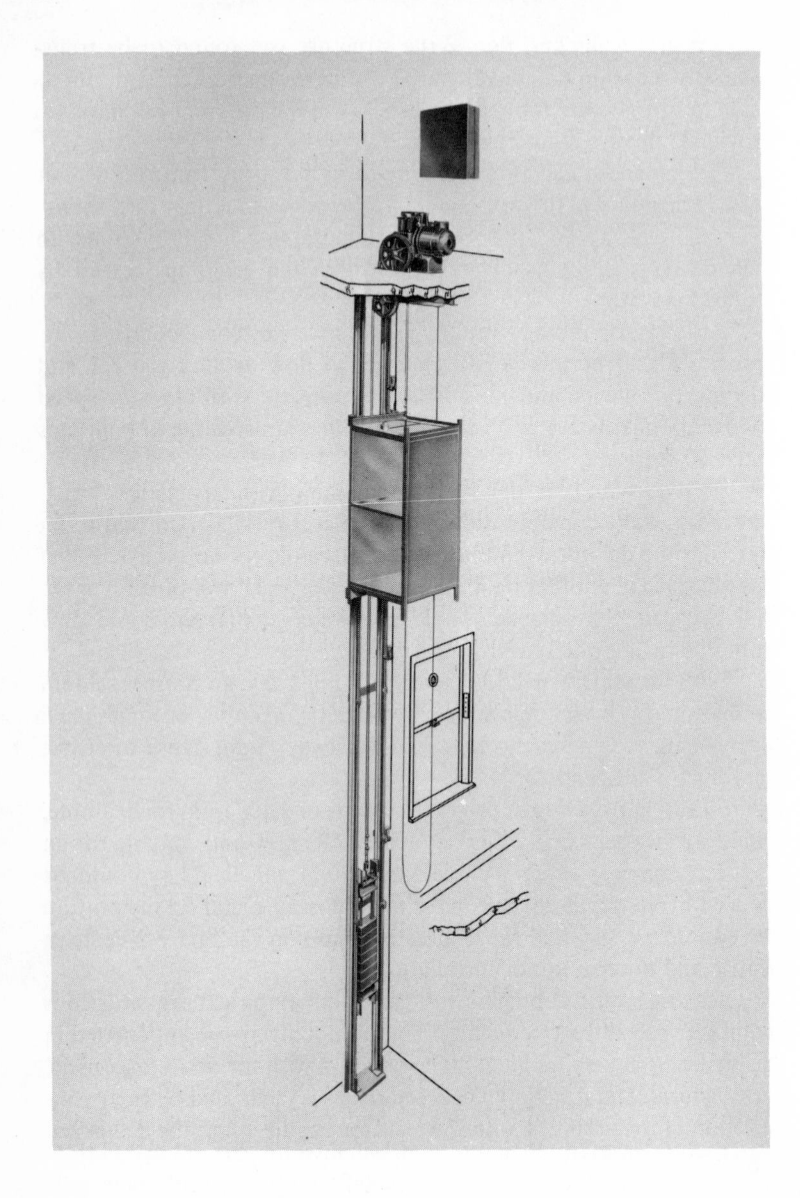

Fig. 4.6. Traction-type electric dumbwaiter with machine-above arrangement.

4.12. *Electric Dumbwaiters*

In hospitals, hotels, libraries, and other buildings, part of the service traffic can advantageously be diverted to automatic electric dumbwaiters. They can handle loads of 50 to 100 lb. at speeds of 45 fpm and faster.

Dumbwaiters not only reduce traffic loads on elevators but also, with special automatic controls and handling equipment, save time and energy for personnel. In planning such installations, architects and engineers integrate dumbwaiters with a building's passenger and freight elevators and with horizontal elements of supply distribution systems.

Electric dumbwaiters for higher rises are built like miniature elevators, with traction-type machine and controller mounted above the hoistway (Fig. 4.6). Low buildings may be served by dumbwaiters with drum-type machines, limited in rise by the length of rope that can be wound on the hoisting drum.

To reduce overhead height, a drum machine can be located at the bottom of the hoistway and the dumbwaiter car equipped with underslung sheaves (Fig. 4.7). An arrangement of this type may be installed as an undercounter dumbwaiter. Guide rails are supported by the building structure, and the controller can be located in any convenient space near the hoistway.

4.13. *Escalators and Moving Walks*

Standardization of escalator design has reduced the relative cost of this form of vertical transportation and contributed to its more extensive application.

Standard escalator widths range from 32 to 48 in. A 32-in. escalator can comfortably carry an adult and child side by side on each step, while the 48-in. escalator can similarly accommodate two adults. Escalators move at 90 or 120 fpm and are reversible.

A modern escalator (Fig. 4.8) is built around the escalator truss, the structural steel frame that becomes an integral part of the building. Factory-fabricated trusses reduce installation cost, and their self-contained driving machinery eliminates separate machine rooms.

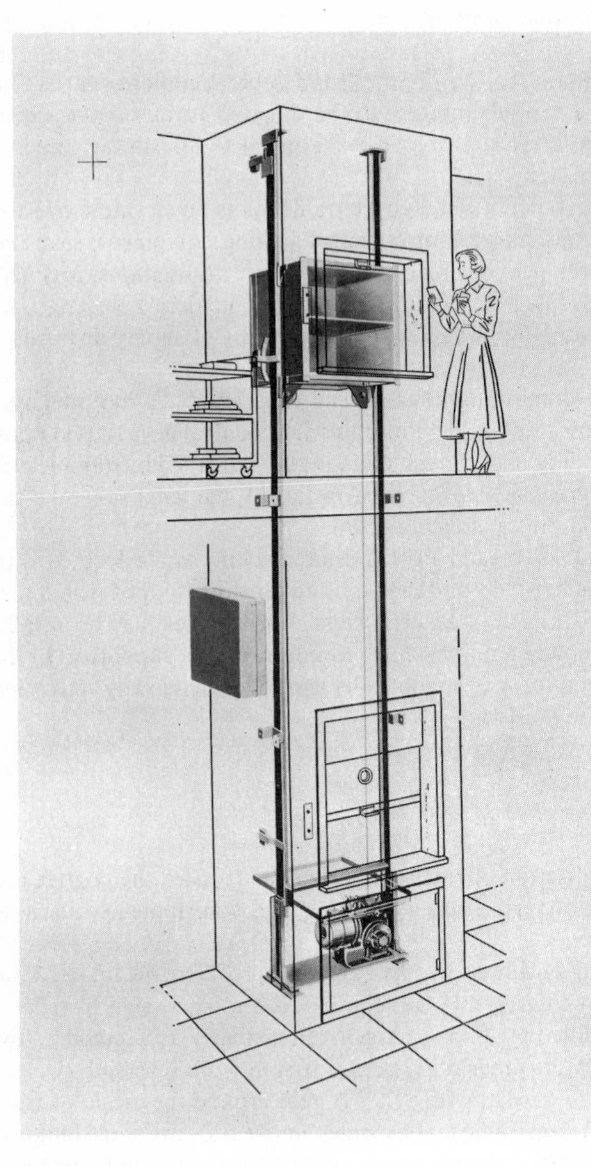

Fig. 4.7. Electric dumbwaiter with drum machine at bottom of hoistway.

The truss contains or supports the treadway mechanism, the interior passenger way, the landing plates, balustrades, handrails, and decks. Modern units have steps with safety cleated risers, semicircular extended newels, pinch-proof handrail openings, replaceable comb teeth sections, and narrow-gauge step treads with safety demarcation lines.

Fig. 4.8. Typical escalator, cut away to show steps and driving machine in truss.

Wellways through the floors of a building accommodate the escalator trusses. Their location depends on whether both units of a pair are placed between two rows of columns or one escalator on either side of a column. The first plan saves space and limits escalator construction work to fewer bays although, if the bays are not wide enough, traffic around the escalators may not move freely. The second plan may simplify framing and leave more space for circulation, but at the expense of higher installation cost. Manufacturers' layout drawings and dimensions guide planning the wellway and supports for each truss.

Although escalator treads are horizontal when they carry passengers, the treadway moves upwards or downward at an angle of 30° or 35° formed by the truss and a horizontal plane. To transport people horizontally or on inclines up to 15°, moving walks (Fig. 4.9) are gaining favor.

Except that it has a treadway with steps only and no risers, a moving walk resembles an escalator in general design and operation. Like escalators, walks are available in 32 and 48-in. widths and move at speeds of 90 and 120 fpm. Moving walks are proving of especial value at air terminals and in shopping centers and other facilities where streams of people must be moved across considerable distances rapidly and comfortably.

Fig. 4.9. Moving walk installation for approach to exhibition building.

Striking visual effects have been achieved in integrating escalators and moving walks, which may have transparent balustrades, with building interiors and exteriors (Section 11.14). Besides their primary function of transportation, such installations may also dramatically expose buildings and their contents to the vision of passengers being carried through them.

Chapter 5

Automatic Control Systems

Efficiency of elevator driving machinery, especially of the gearless traction type (Section 4.2) reached a high level many years ago. Yet recent progress in improving the effectiveness and economy of elevator service has been dramatic. Much of the gain stems from developments in automatic control systems (Fig. 5.1) that more fully

Fig. 5.1 Part of elevator lobby of 60-story Chase Manhattan Bank Building in New York. Handling heavy traffic most of the day, elevators in this and other modern buildings operate completely automatically. Cars of each group are also coordinated by automatic supervisory systems to keep service adjusted to traffic.

utilize the performance capabilities of the basic elevator plant to meet the diverse transportation needs of elevator users.

5.1. *Trend to Automation*

Service quantity and quality depend on the number, size, and speed of elevators and the effectiveness with which they are controlled.

Increasing costs and decreasing availability and dependability of building-service labor in recent decades intensified demands for the completely automatic operation of elevators, even in buildings with heavy traffic. At the same time, equipment for elevator automation was being improved in capability and reliability and reduced in relative cost. As a result nearly all elevators installed since 1950 have been highly automated and older installations have been modernized to operate without attendants.

Owners and managers of office buildings in North American cities report that automation tends to reduce operating costs by about half, compared with manually controlled elevators. Major savings come from salaries and the staff of elevator attendants and starters necessary for full-time service. Elimination of fringe benefits, employment taxes, expenses for uniforms, and administrative costs are further economies creditable to operation without attendants.

Automation also attains standards of elevator service impossible with manual operation. A system of automatic controls properly engineered to a building's traffic requirements keeps its elevators performing at peak efficiency, unhampered by the human limitations of attendants and starters. Car and door operation is speeded, saving seconds at every stop for every elevator. Gains are cumulative, enabling elevators to complete trips in less time and resulting in faster, more frequent service for passengers and greater carrying capacity for the installation.

A control system that achieves maximum elevator performance provides the necessary transportation with minimum investment in basic plant, the costliest part of the installation. More service with fewer cars may also save hoistway and corridor space at each floor and require less construction to accommodate elevator machinery.

In new construction more intensive utilization of elevator plant through effective automation permits planning a building with a higher ratio of net to gross space. In modernization, hoistway space may be recaptured for other purposes.

Automatic control of a building's elevators may embrace three distinct yet interrelated functions: power control, elevator operation, and group supervision. Various methods and systems are used to perform each of these basic functions.

Power control refers to means for starting, stopping, accelerating, decelerating, and reversing the elevator drive machinery. Even a manually operated elevator requires at least a rudimentary system for power control.

Elevator operation embraces means for automatically regulating the movement and position of the car in response to passenger calls. A closely related function, in most automatic elevators today, is the operation of car and hoistway doors.

Group supervision, finally, comes into play when two or more elevators operate as a group (Fig. 5.1). The supervisory system automatically coordinates distribution of the several cars to match service of the entire group to the volume and distribution of traffic.

5.2. *Power Control Systems*

Depending on the precision and smoothness of elevator operation required, power control may be accomplished by variable voltage or resistance-type systems.

Variable voltage systems control elevator acceleration, running speed, and deceleration with the greatest smoothness and accuracy, regardless of load. This type of control may be applied to elevators traveling at speeds of 100 fpm to 1,800 fpm or faster.

For variable voltage control, the drive motor of each elevator machine is energized by its own motor-generator set, controlled in turn by a controller and associated equipment. The elevator is slowed down by its driving motor's dynamic braking action rather than by the brake, which acts normally only to hold the car at a floor. As systems are developed to achieve dynamic braking with silicon control rectifiers, or thyristors, they may replace motor-generators for speed control.

In present variable voltage systems the speed for the elevator drive motor is directly proportional to the voltage impressed on its armature by the generator. Voltage is smoothly varied by the action of the controller changing the strength of the generator field, resulting in smooth, virtually stepless operation of the elevator under all conditions.

Rapid starts, higher speeds, and accurate stops are achieved without passenger discomfort and traffic-handling capacity is increased by decreasing round-trip time. Predetermined speeds are uniformly maintained without slowdowns due to heavy loads during traffic peaks. Gradual acceleration materially reduces wear and tear.

Resistance control is used for slower elevators, of the geared and hydraulic types. The driving motor, a single-speed or two-speed a-c unit, is placed across the line with a step or two of resistance. To stop the elevator, a controller shuts off power and applies a brake. Stopping accuracy varies with load.

5.3. *Elevator Operating Systems*

Selection of a system for automatically controlling elevator operation depends primarily on the intensity of elevator use expected and the quality of service desired. In order of their capability for satisfying increasingly intensive service demands, operating systems are of the single automatic, collective, and group supervisory types.

As its name implies, single automatic operation permits an elevator to serve only one call at a time. Each landing has a single call button, which a passenger presses to summon the elevator.

In the car he presses a button for the floor he wishes to reach, to which the elevator travels directly without stopping en route for other hall or car calls. This method of automatic operation was introduced as early as 1894 and is still used for elevators in private residences and for freight elevators which usually carry only a single load at a time.

With collective or group supervisory operation, each elevator can answer many calls on each trip. Intermediate landings usually have both up and down buttons, and the car has numbered floor buttons.

To use the elevator, passengers press or touch the up or down landing buttons and the floor button in the car. Landing and car buttons are connected to circuits which electronically "remember" all calls and control the movement of the elevator to answer them.

As the elevator moves up in the hoistway, it stops automatically as it reaches each floor for which someone has pressed a button. When the elevator has answered the highest call, it automatically reverses and then stops at floors for which there are down calls.

Collective operation is used for installations of one or two elevators handling light traffic; group supervisory systems control two or more elevators with heavier traffic.

Modern collective or group supervisory operation includes two-way automatic leveling. With modern equipment an elevator is automatically brought to a stop within one-half inch or less of floor level, whether the car is lightly or heavily loaded. Automatic leveling prevents many tripping accidents at elevator entrances.

5.4. *Door Operation and Reversal*

Power-operated car and hoistway doors are preferred in modern automatic elevators, with car and door operation and leveling closely coordinated to save time from one stop to the next. Suitable door reversal devices protect passengers while reducing the time required to close the doors, letting an elevator leave sooner after each stop.

Where traffic is intermittent, door reversal may be initiated by rubber safety edges mounted on the leading edge of the car doors. If the rubber edge touches a passenger as the doors are closing, car and hoistway doors automatically reverse and reopen.

For added protection, the rubber edge may be supplemented by a light source and photocell mounted on opposite sides of the car entrance. A passenger walking through the entrance interrupts the light beam, operating the door reversal mechanism.

Electronic detectors for door reversal (Fig. 5.2) are among the developments that made "operatorless" elevators practical for intensive service. This radarlike device safely speeds door operation by permitting doors to continue closing, without reversing, unless they come within a short distance of a passenger, when the detector stops and reverses the doors before they touch him. If the person persists in standing in the path of the doors, associated "nudging" circuits close the doors slowly but surely to ease him safely aside so that the elevator can continue on its way.

Fig. 5.2. Electronic detector door-reversal device permits power-operated doors of automatic elevator to close quickly without interfering with passengers entering or leaving car. If passenger enters "zone of detection" extending along, across and in front of car and hoistway door, this proximity device automatically reverses and reopens doors.

Fig. 5.3. Central station dispatching system instructs dumbwaiter to stop at several floors and, after making its highest delivery, to return to the central station, all automatically. Door opening and closing may be completely automatic but in most installations a person must close doors manually to permit dumbwaiter to start on its next trip.

Limited door reversal, under the control of electronic detectors with special additional circuitry, further speeds elevator transportation. Doors reopen, not all the way, but only wide enough to let a passenger through, and then close again so that the elevator can depart promptly.

5.5. *Dumbwaiter Operation*

Whether a dumbwaiter is used intermittently or intensively influences the selection of automatic controls. For intermittent use a dumbwaiter serving two floors may have a "call and send" control system and, for three or more floors, a "multibutton" system. Dumbwaiters for intensive service are controlled by "central station dispatching" systems (Fig. 5.3).

For "call and send" dumbwaiter operation, each of the two landings served has a fixture with up and down buttons. A person momentarily presses a button to call the car to his landing or send it to the other.

For multibutton operation the fixture at each landing has one "call" button, and several "send" buttons numbered for the other landings. To call the car to his landing a user presses the "call" button and, to send it to another landing, presses the button numbered for that landing.

Whereas multibutton dumbwaiter operation is similar to single automatic elevator operation and permits sending the car to one destination at a time, "central station dispatching" for a dumbwaiter is comparable to collective operation for an elevator.

At the landing from which the dumbwaiter is centrally dispatched, such as the central sterile supply room of a hospital, a landing fixture has buttons numbered for each of the other floors served. A user places loads for several floors in the car and presses the buttons for those floors. The dumbwaiter travels to each of the desired floors and then automatically returns to the central station.

Recent developments permit a dumbwaiter to be loaded as well as unloaded automatically, picking up and ejecting trays at counter-height entrances or carts at floor level. Installations of this type help to automate the distribution of food and supplies in hospitals, mail and memoranda in office buildings, and smaller parts and materials in industrial plants.

5.6. *Elevator Group Supervision*

Quality and quantity of service depend on coordination of the entire group of elevators as well as on control of individual cars. Automatic group supervision keeps the operation and distribution of all cars adjusted to the prevailing pattern of traffic intensity and distribution, so that elevators answer passenger calls promptly.

With efficient supervision, a passenger who just misses an elevator during a period of intensive traffic need not wait the full round-trip time but, on the average, only the round-trip time divided by the number of cars in the group. Service of all elevators is equally available to all passengers, on all floors, and the result is minimum average waiting time.

Elevators controlled by a modern group supervisory system work as a team to handle a broad range of service demands with economy of time and travel. The system keeps elevators distributed in such a way that they are always ready to answer calls as promptly as possible.

5.7. *Systems for Moderate Traffic*

For efficient two-car service a relatively simple form of supervision automatically coordinates operation of a pair of collective elevators. Supervisory circuits assign each call to the car in better position to answer it and prevent both cars from answering the same call. One elevator usually provides service from the entrance floor up into the building while the other serves passengers traveling between upper floors and down to the entrance floor.

To supervise a group of three or more elevators, for efficient service during periods of moderate traffic, a completely automatic system known as multiple zoning (Fig. 5.4) has been developed. The system controls the spacing of elevators throughout a building so a passenger will find a car at or near his floor, minimizing needless elevator travel and reducing passenger waiting time.

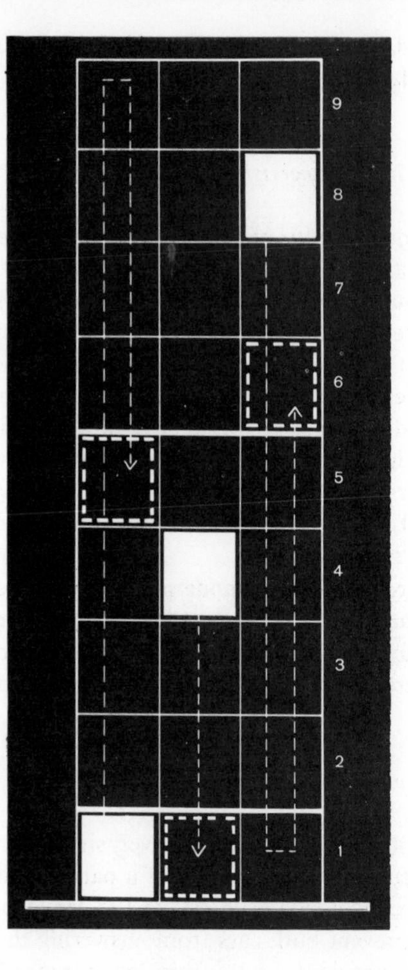

Fig. 5.4. Multiple zoning automatically controls distribution of elevators throughout a building so that a car is near a call. Floors are separated into zones usually as many as there are cars, and elevators parked (solid-line squares) so that each car "covers" a zone. Cars answer passenger calls from their respective zones, then park (broken-line squares) so that each zone is again covered.

Floors are separated into zones, usually as many as there are cars, and each car is assigned to a zone. An elevator answers all calls from the zone it is "covering" at the time. After answering calls, the elevator parks in an unoccupied zone.

Information on passenger demand at all floors and in all elevators is continuously monitored by sensors—the buttons at landings and in cars, and load weighers (Section 5.8)—and fed to a special-purpose computer in the machine room. Processing elevator traffic data in real time, the computer can change zone assignments to meet actual and anticipated demands for flexible, responsive group supervision. With high-speed elevators coordinated by multiple zoning, passengers not only enjoy service with little or no waiting but are also carried to their floors as directly as possible.

5.8. *Systems for Intensive Traffic*

During periods of heavier traffic, when people call for elevators almost continuously, each group must provide more intensive service. For these conditions, group supervisory systems incorporate circuitry designed to operate cars continuously, automatically matching the distribution of elevators to the distribution of calls so that passengers throughout the building enjoy prompt service. Group performance should be controlled to provide equal service in both directions or any ratio of up to down service that traffic patterns require.

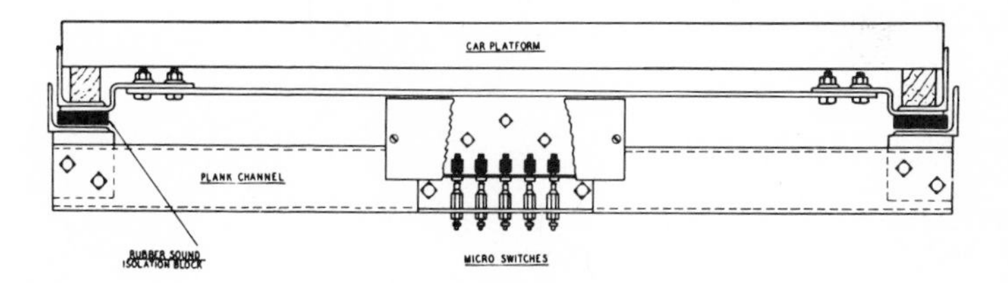

Fig. 5.5. Load weigher automatically detects when elevator is full, directs it to start ahead of schedule or to bypass hall calls until car has room for more passengers. Keeping elevators from unnecessarily waiting or stopping for passengers for whom the car has no space, load weighing helps distribute elevator group capacity.

When up or down elevator traffic attains peak levels, as at the start or close of the working day in an office building, the supervisory system can automatically detect the situation and operate elevators to maximize carrying capacity in the desired direction. Data from sensitive load weighers (Fig. 5.5) for each car may modify operation during traffic peaks. These sensing elements electronically weigh the total passenger load in each car and signal the supervisory computer accordingly.

Chapter 6

Commercial Buildings

In the late nineteenth century, office buildings of greater and greater height began to punctuate urban skylines, first in North America and then in other parts of the world. These "skyscrapers" were made possible by steel frame construction and high-speed elevators and profitable by the growth of commercial activity centered in the business districts of bustling cities.

For a time after World War II, developments in electronic communication and highway transportation seemed to foreshadow a trend away from high-rise office buildings concentrated in compact centers. Then, as businessmen became aware that the reproduced image or voice could not replace the face-to-face encounter, towers in increasing numbers began rising to new heights in cities of the advanced nations.

As a corrective for urban congestion, vertical design can shift much of the traffic load from overburdened streets to capacious systems of high-speed elevators. So that cities may benefit more fully from this principle, the Regional Plan Association of New York has proposed clusters of high-rise buildings over subsurface transit stations, with "access trees" of elevators and escalators (Fig. 6.1) linking stations directly to buildings. Elements of this concept have already become a reality in individual buildings and multibuilding developments in a number of cities.

6.1. *Elevator Service Requirements*

Whether an office building is part of a unified center or is located in the more conventional business district, an economically successful building must provide vertical transportation to satisfy the needs of its businessman occupants and visitors.

Business people have long held that "Time is money." They not only need elevator service to reach offices high in downtown towers, but also insist on service that takes them there without delay.

To be available to transact business with each other, businessmen in the same line or locality tend to keep the same hours. Many of the occupants in an office building therefore come to work and leave for home at about the same times each day, imposing highly concentrated demands on the building's elevators.

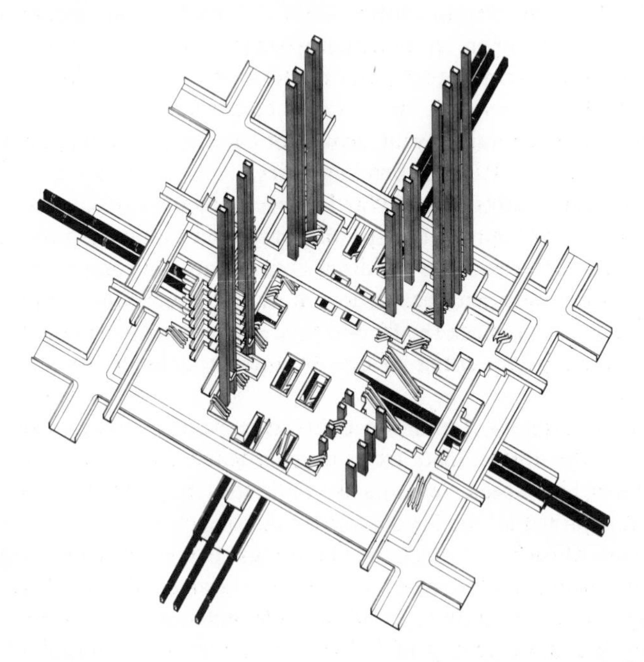

Fig. 6.1. Midcity convenience without congestion is envisioned by the Regional Plan Association of New York "access tree" concept, in which elevators and escalators directly link transit stations to office towers clustered above.

For these reasons, systematic study of vertical traffic and formulation of elevator planning principles were first applied to office buildings.

Traffic in most office buildings and many hotels and motels tends to be concentrated in well-defined periods, or peaks, of relatively limited duration, as outlined in Sections 6.2 through 6.10. Large retail stores and other types of commercial buildings, reviewed in

Sections 6.11 to 6.14, experience heavy vertical traffic persisting for longer periods, perhaps several hours.

6.2. *Occupancy and Vertical Traffic*

In the central business district of the modern commercial city, many office buildings are of the diversified-tenancy type, and a few are classed as single-occupancy and others as professional office buildings.

As the name implies, the diversified building accommodates a large number of different tenants in various lines of business. This type of occupancy creates distinctive requirements for vertical transportation since the office hours of all tenants, although similar, are not identical.

Service demand reaches a peak during the morning and evening periods when most occupants are going to work or leaving for the day (see Fig. 2.3), but the peaks are more extended and less intense than in a building occupied entirely by a single organization. Traffic is mostly up during the morning peak, down in the evening. Another period of intensive traffic occurs around midday as people leave for and return from lunch. Elevators must be able to handle the peaks and also provide fast, frequent service all day long.

For much of the morning and afternoon, traffic may be moderate in volume compared with the opening, closing, and midday periods. After working hours, traffic becomes light and may even cease entirely for periods.

In determining the capacity of a building's vertical transportation system, the morning inrush peak makes the most severe demands on equipment. Tenants desire that their employees reach their offices by starting time, yet passengers are reluctant to load elevators or escalators as fully as during the evening, going-home peak.

Elevators may make many stops on the way up, reducing effective handling ability as compared with the down peak, when elevators may fill near the top and descend directly to the main floor. As a result, if elevators have sufficient capacity to handle traffic during the busiest 5 minutes of the morning peak, they can usually meet the heaviest demands of any other period of the day.

Peak 5-minute traffic during the critical inrush period may range from 10 to 15 percent of the total population above the street floor of a diversified-tenancy building (Table 6.1). A rule of thumb for approximating the population of a proposed building has been an average of one person per 100 sq. ft. of net floor area. As automatic data processing systems perform more of the routine tasks in business offices, a larger proportion of staffs will consist of executive and professional personnel with larger space allowances than the traditional average.

Table 6.1
Office building five-minute peak traffic as percent of building population

Diversified-tenancy building occupied largely by executive or professional personnel	12–15%
Average diversified-tenancy building in area served by good horizontal transportation	15–20%
Single-occupancy building (*e.g.* insurance company) with strict employee discipline	20–30%

Floor-by-floor distribution of demand is influenced by characteristics of building design and occupancy. Entrances from two levels and dining facilities on upper or lower floors, for instance, affect traffic distribution. If one tenant has offices on several floors, elevators may have to serve appreciable interfloor traffic as well as traffic between the main and upper floors.

6.3. *Standards of Service Quality*

Elevators and escalators must be able to provide service satisfactory in quality as well as in quantity (Section 2.6). Passengers probably expect transportation service of higher quality, in terms of frequency and speed of service, in office buildings than in those of most other types.

As employment costs continue to rise, office-building owners and tenants find prompt service with a minimum of waiting and riding time increasingly valuable. By gaining a few seconds for each passenger on every trip, effective vertical transportation can, in the course of a year, save thousands of costly man-hours for all the people in a building.

In a first-class downtown office building, passengers usually become impatient if they must wait longer than 30 seconds for an elevator. For service of acceptable quality, elevators should therefore answer calls within 30 seconds and, for preferred quality, 25 seconds or less. Expectations of service frequency and speed vary somewhat from one locality to another.

Consistency of service is also important, since unexpectedly long waits are annoying. Once a passenger has entered an elevator, he prefers to reach his floor as quickly and with as few intermediate stops as possible.

6.4. *Planning the Plant*

Each vertical transportation system is designed to handle traffic of the volume and distribution expected in a particular building and to provide service of the desired quality. Factors that influence the location of an elevator or escalator plant, the arrangement of its component units, and their operating specifications in any type of building (Sections 3.4 to 3.8) also pertain in the case of an office building. Other factors that apply more specifically to the latter type are outlined below.

Because of the tallness of the typical diversified-tenancy office building and the preponderance of traffic between street level and upper floors, high-speed elevators are usually the primary means of transportation.

If considerable traffic is also anticipated between the main floor and levels immediately above or below it, escalators or separate shuttle elevators may supplement the main elevators. Such arrangements, often used for access to basement dining, parking, or transit facilities or mezzanine banking or retail facilities (Figs. 6.2 and 6.3) avoid the necessity of more than one lower terminal for the main elevators, a situation which could impair service.

In addition to escalators for its lower floors and elevators for those higher up, a fairly conventional plan, a building in Dallas, Texas, also has escalators between the forty-ninth and fiftieth floors. The latter escalators were necessitated by legal height limitations to safeguard approaches to a nearby airport. Escalators rather than elevators to the highest floor eliminated a rooftop penthouse for

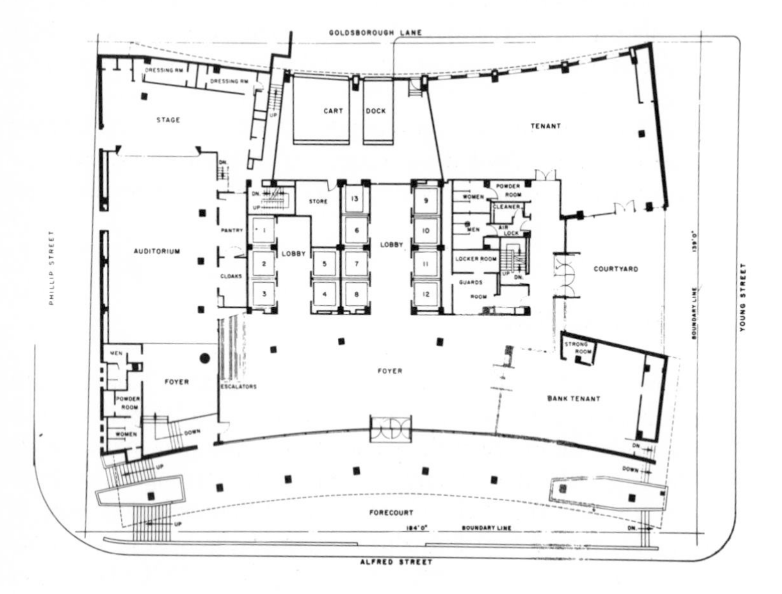

Fig. 6.2. Australian Mutual Provident Society Building, Sydney; Peddle, Thorp & Walker, architect. Street-floor plan shows escalators to floor immediately above, passenger elevators (1–12) and service elevator (13) to all upper floors.

elevator machinery which would have exceeded the limit on total height.

Vertical transportation usually forms part of a building's utility core, strategically located to save steps for people (Fig. 6.4). At street level, elevators and escalators should be convenient to the main entrance.

Instead of a narrow corridor around all four sides of the elevator core, a wider corridor across the middle may in some buildings improve circulation and reduce total space needed for this purpose. In others, to gain large, open floor areas, vertical transportation and other utilities may be installed in a service tower beside the main structure.

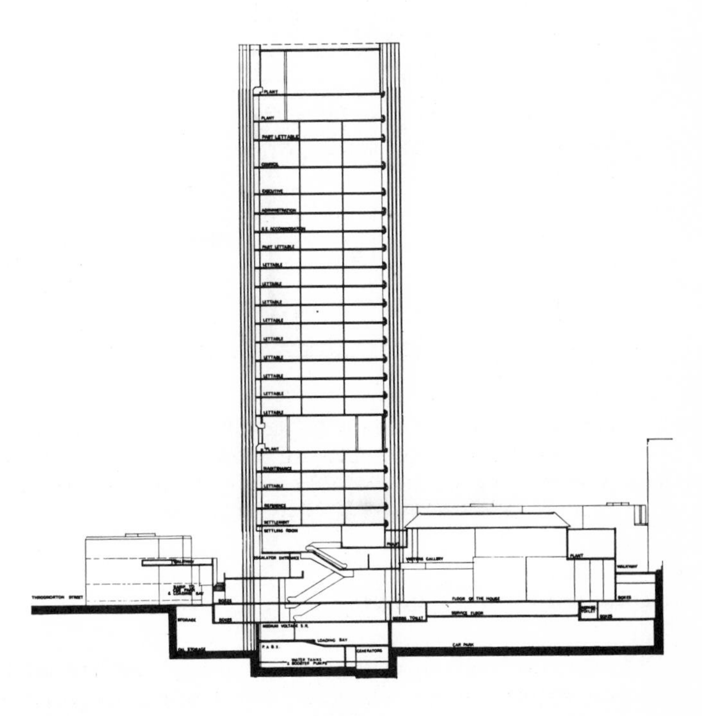

Fig. 6.3. Redevelopment of London Stock Exchange. Llewlyn-Davies, Weeks & Partners, and Fitzroy Robinson & Partners, architects; Ove Arup & Partners, consulting engineer; Trollope & Colls Limited, general contractor. Office floors in 26-story tower block (center) are served by eight passenger elevators, in two groups of four cars each, and two service cars. Six escalators in base of tower carry brokers and clerks between levels of Settling Room, the trading floor of the Market Block (right), and a lower level of "boxes" or cubicles for brokers.

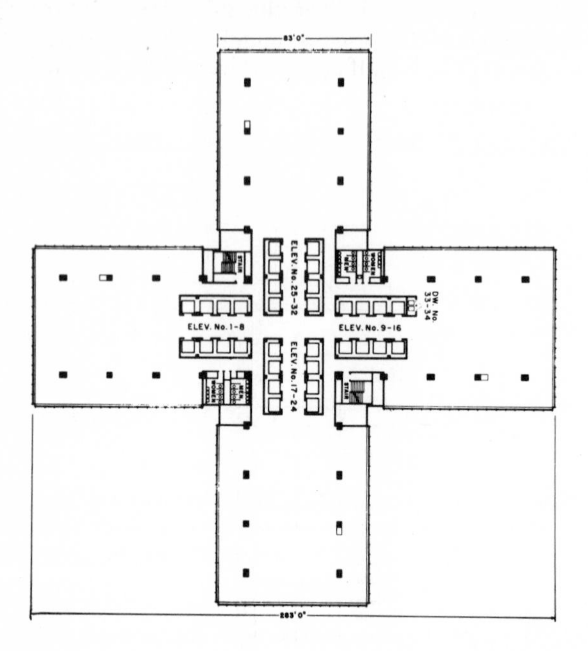

Fig. 6.4. The Royal Bank of Canada Building, Place Ville-Marie, Montreal; I. M. Pei & Associates, architect. Forming the building's vertical core is a system of 32 elevators (shown in typical floor plan) and 14 escalators serving five lower levels. Elevators 25–32 are service cars.

6.5. *Local and Express Elevators*

As a rule, diversified-tenancy office buildings are large and tall enough to require several elevators. For greatest frequency and continuity of service, cars are located together and operated as groups.

Office buildings of twenty or more stories often have enough elevators for two or more groups. Separate high-rise, low-rise, and, possibly, intermediate-rise groups with sufficient cars in each can improve the quantity and quality of service to all floors and may, at the same time, reduce total elevator installation and building construction costs (Section 3.5).

Elevator groups in office buildings may have four, six, or eight cars, smaller groups often being favored for local service and larger groups for express (Fig. 6.5). If appreciable traffic is expected between

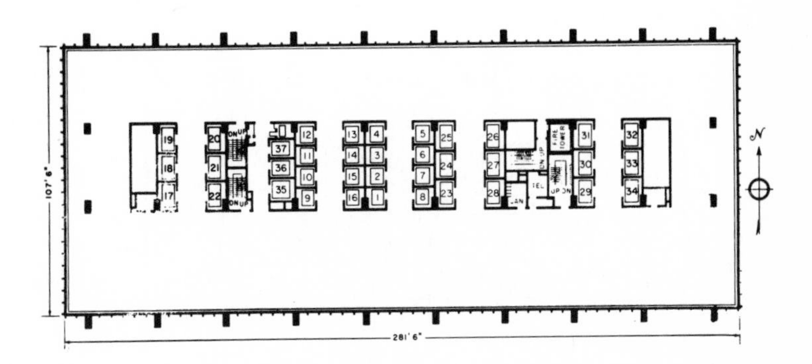

Fig. 6.5. Chase Manhattan Building, New York; Skidmore, Owings and Merrill, architect. Local and express elevators serving upper floors are arranged in six- and eight-car groups. Elevators 35, 36, and 37 (car numbers not shown) are service cars.

floors served by a local group and the next higher express group, the system design may include a transfer floor, usually the highest stop for the local elevators and the first stop, above the main floor, for the express group.

6.6. *Size and Speed*

In smaller office buildings, passenger elevators of 2,500 or 3,000 lb. capacity may suffice. Taller buildings with heavier traffic are better served by 3,500-lb. or 4,000-lb. elevators (Table 6.2), with wide, shallow cars and center-opening doors to speed passenger transfer.

Travel speed, the other major element in handling capacity, may range from 250 fpm to 1,800 fpm or faster (Table 6.3). Higher speeds are attained only on longer express runs, since acceleration and deceleration are limited to rates that passengers find comfortable.

Gearless traction machines attain the speeds required for elevators in taller office buildings. This type of drive is also preferred

Table 6.2
Elevator sizes for commercial buildings

Rated capacity		Platform		Hoistway clear		Doors (standard ht. 7′ 0″) clear opening
pounds	passengers	(w) width	(d) depth	(W) width	(D) depth	
2,000	13	6′ 4″	4′ 5″	7′ 8″	5′ 9″	3′ 0″
2,500	16	7′ 0″	5′ 0″	8′ 4″	6′ 4″	3′ 6″
3,000	20	7′ 0″	5′ 6″	8′ 4″	6′ 10″	3′ 6″
3,500	23	7′ 0″	6′ 2″	8′ 4″	7′ 6″	3′ 6″
3,500	23	8′ 0″	5′ 6″	9′ 5″	6′ 10″	4′ 6″
4,000	27	8′ 0″	6′ 2″	9′ 4″	7′ 6″	4′ 0″

Table 6.3
Elevator speeds for office buildings

Floors	Passenger (fpm)	Service (fpm)
2–5	250–400	200
5–10	350–500	300
10–15	500–700	350–500
15–25	700–800	500
25–35	800–1,000	500
35–45	1,000–1,200	700–800
45–60	1,200–1,600	800–1,000
60 or more	1,800	1,000

for its acceleration and deceleration characteristics, quiet operation, and long service life.

Elevator machinery is mounted directly over the hoistway (Fig. 6.6) in a machine room with dimensions depending largely on the number and speed of elevators. Hoistway dimensions (Table 6.2) depend on elevator size, and pit depths and overhead heights (Table 6.4) are governed by elevator speeds and local building codes.

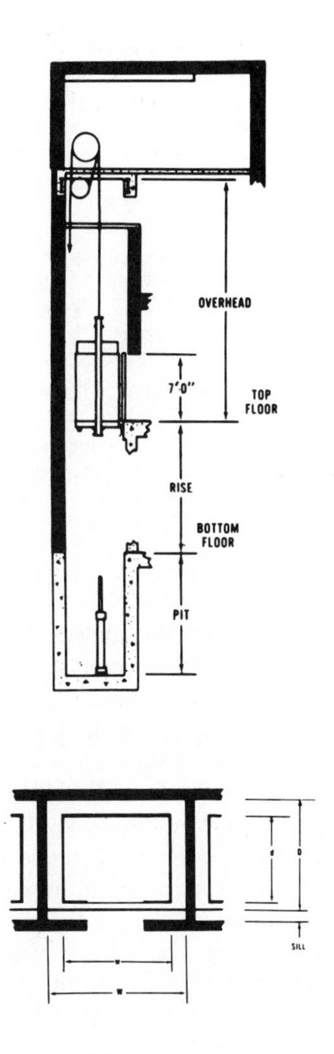

Fig. 6.6 Office-building elevator section and plan. See Table 6.2 for hoistway dimensions and Table 6.4 for pit depths and overhead heights. Sill is 4 in. for single-slide or center-opening doors, $5\frac{1}{2}$ in. for two-speed doors.

Table 6.4
Minimum pit depths and overhead heights for certain elevator speeds and sizes

Speed (fpm)	Capacity (lb.) 2,500 pit	2,500 o. ht.	3,000 pit	3,000 o. ht.	3,500 pit	3,500 o. ht.	4,000 pit	4,000 o. ht.
250–300	5′ 2″	16′ 2″	5′ 2″	16′ 2″	6′ 9″	15′ 8″	—	—
350	5′ 2″	16′ 2″	5′ 2″	16′ 2″	—	—	—	—
400	—	—	—	—	8′ 0″	22′ 2″	—	—
500	8′ 2″	17′ 10″	8′ 4″	18′ 8″	8′ 6″	22′ 2″	8′ 8″	23′ 6″
600	9′ 2″	23′ 6″	9′ 2″	23′ 6″	—	—	—	—
700	9′ 10″	24′ 6″	9′ 10″	24′ 6″	—	—	—	—

Pit depths are measured from lower landing floor to pit floor, overhead heights from upper landing floor to top of machine beam supports.

6.7. *Automatic Control*

High standards of service demanded in office buildings are achieved with elevators of sufficient number, size, and speed, controlled by automatic systems that realize the full potential performance of which an installation is capable (Chapter 5).

For smooth, time-saving acceleration and deceleration, elevator speed should be controlled by the variable voltage method. Speed and convenience of service demand automatic leveling synchronized with the opening of power-operated doors. Accurate leveling also reduces tripping hazards to a minimum, a significant contribution to safety in a building with elevators accessible to the public at large.

In smaller towns or outlying neighborhoods, low-rise office buildings with light vertical traffic may be adequately served by automatic elevators with collective or duplex collective operation (Sections 5.3 and 5.7). Taller, busier buildings require more advanced systems of automatic elevator operation and group supervision capable of handling the traffic encountered and providing service of the standard desired (Sections 5.6 and 5.8). A well-engineered control system helps achieve a primary objective of office-building elevator service, carrying all passengers to their floors with least loss of time waiting for and riding on elevators (Section 13.3).

6.8. *Service Elevators*

Although tenants and visitors see the passenger elevators far oftener than the service cars, the latter are also essential to successful building operation.

Incoming material and outgoing trash in an office building are estimated to total 150 to 200 lb. per week per employee. A 250,000 sq. ft. building with 2,000 to 2,500 occupants would have some 250 tons of material to be moved to and from upper floors every week. That would keep a 4,000-lb. service elevator busy all week long, assuming that the elevator makes a round trip every 10 minutes and is loaded, on the average, to half its capacity.

Service-elevator traffic includes mail, supplies, furnishings, carts for coffee or cleaning, and even building equipment and materials. Space alteration and renovation is an almost continuous process in many buildings, making it necessary to carry masonry, mortar, and partitions from floor to floor.

Much of this material must be moved at times when the passenger elevators are otherwise busy. At present and prospective labor rates, to delay the moving of materials until a passenger elevator can be spared for the purpose may cost more, all costs considered, than installing separate service elevators.

Service elevators of 3,500 or 4,000 lb. capacity, with platform dimensions similar to those of passenger elevators, are satisfactory in smaller buildings. Larger, taller buildings may require elevators of 5,000 lb., 6,000 lb. or more to handle massive equipment like air-conditioning system pumps and power distribution transformers. Possible failure of such equipment may necessitate a capacious elevator to remove and replace the entire unit for prompt restoration of the affected service.

Entrance dimensions depend on the size of the largest equipment to be carried. For wider openings than standard center-opening doors afford, the two-speed sliding or two-speed center-opening type may be used. If an elevator handles considerable freight traffic but no passengers, local codes may permit vertical biparting entrances, with openings almost as wide as the car platform itself. Car ceiling and entrance opening heights usually exceed those of passenger elevators.

For occasional handling of especially heavy loads a service elevator may have provisions for "safelift" operation. They include high-rated driving machinery, special locking arrangements, and a counterweight frame to which extra weights can be added to permit temporarily increasing the elevator's normal lifting capacity by one-third.

Loading and unloading account for a greater proportion of round-trip time than for passenger elevators. Since running time is less important in the service-elevator operating cycle, travel speeds may be slower than for passenger elevators of comparable rises (Table 6.3). Automatic control systems are similar, except that the service car may also include optional attendant operation for purposes of administration and security.

Service elevators must be located conveniently to a lower-level service entrance or loading dock and to upper-floor service-traffic destinations like kitchens, mechanical areas, and storage facilities. In some smaller buildings, service elevators are often adjacent to a group of passenger cars and operate as extra passenger elevators to handle peaks in the latter kind of traffic. During such "help out" operation, the service-car control system functions as part of the group supervisory system for the passenger elevators.

Flexibility of use may be facilitated by a service elevator with two entrances, front and rear or front and side. Front entrances, opening on passenger elevator corridors, operate only when the service elevator is lending its capacity to the passenger group. For service use the other entrances operate, the changeover being controlled by a keyswitch.

6.9. *Single-Occupancy Buildings*

If one organization occupies an entire building and most of the staff works during the same hours, vertical traffic peaks are appreciably more severe than in diversified-tenancy buildings (Table 6.1).

In some single-occupancy buildings, management can stagger

office hours to reduce the intensity of traffic peaks while extending their duration. In other cases, carrying capacity is maximized by "spotting" elevators during traffic peaks.

At these times each elevator in a group operates between the lobby and only one or two designated upper floors. Since elevators shuttle back and forth with few if any intermediate stops, they complete more round trips and carry more passengers in a given period of time. After the peak period, elevators resume normal group operation.

Where elevators can be spotted, they may have oversize cars of 5,000 or 6,000 lb. capacity. Lobbies must be planned with space to accommodate people queueing up for designated cars.

Interfloor traffic may be heavier in a single-occupancy than in a diversified-tenancy building. Owner-occupied buildings up to eight or nine floors high may use escalators as the primary means of vertical transportation supplemented by one or more elevators for handicapped persons or service traffic.

In buildings of moderate height, escalators often prove superior not only for interfloor movement but also for peak traffic. Reversing the proper escalators (Section 3.2) provides all up service during the morning inrush, all down service during the evening outflow, and up and down service for other periods of the day.

If only persons in able physical condition use escalators in a single-occupancy building, they can move at 120 rather than 90 fpm, increasing carrying capacity and service speed by one-third.

6.10. *Professional Buildings*

Sometimes called "medical arts" buildings, professional buildings are occupied by physicians, dentists, and related practitioners and technicians. Whether located in a downtown business district or suburban shopping center, these buildings are accessible to large populations who use the services they house.

Morning arrival and evening departure of employees, which dominate traffic patterns of other office buildings, are relatively less significant in the professional building. Its nurses, receptionists, secretaries, and technicians keep conventional office hours, but physicians and other professionals come and go at various times of the day.

Throughout the day most of the vertical traffic in the typical professional building consists not of occupants but of visitors: patients accompanied by families and friends. This traffic, according to one estimate, may average 8 persons up and an equal number down, for every office in the building, during every hour of the working day.

The number of offices above the street floor thus provides a basis for calculating the quantity of elevator service required. Service quality is generally satisfactory if elevators operate at intervals of 40 to 50 seconds.

Because many patients are not in the best physical condition, primary reliance for vertical transportation in the professional building must be placed on elevators, possibly supplemented by escalators. Passenger elevators are similar to those in other office buildings with at least one car of the hospital-service type (Section 8.2) for stretcher and wheelchair cases.

6.11. *Commercial Hotels and Motels*

Recent trends in hostelries tend to blur the distinction between hotels and motels. Whether called a hotel or a motel, a modern inn offers its guests convenient, easy parking. Today's motel is as likely as is the hotel to be located downtown close to commercial and cultural attractions. Hostelries appealing primarily to persons traveling on business rather than on pleasure are also being built at airports and near industrial parks.

Such a hotel or motel probably has, in addition to several hundred guest rooms, a sizeable ballroom, meeting room, or restaurant facilities. Public functions in these rooms are attended by large groups of people, many of whom stay elsewhere. To satisfy the demands of guests and visitors, the inn requires vertical transportation comparable in quantity and quality to other busy commercial buildings.

Public-function facilities and entrances from a driveway or garage, usually on lower floors of the hotel or motel, need their own elevators or escalators. Guest-room floors, in turn, have one or more groups of high-speed elevators like those in office buildings. Separate service elevators are also usually necessary, their capacity depending on the load expected from food-service and room-service operation (Fig. 6.7). Separating public-function, guest-room, and service traffic not only speeds handling of each kind but also promotes the comfort of all patrons.

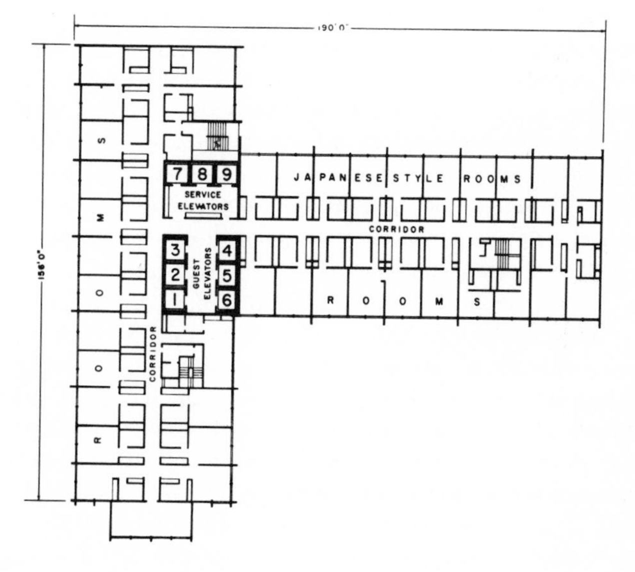

Fig. 6.7. Hong Kong-Hilton Hotel; Palmer & Turner, architect. Guest elevators and service elevators form separate groups near central core. Four escalators (not shown) serve three lower levels.

Compared with vertical traffic peaks in office skyscrapers, those in commercial hotels and motels are of longer duration but less intensity. Early evening is usually the busiest time, as guests check in or out or leave the establishment for evening activities. Elevators or escalators to the public facilities must handle heavy surges of traffic when functions in those rooms begin or end.

Since hotels need elevator service 24 hours a day, 7 days a week, the economics of automation prove extremely attractive. Guests have become accustomed to automatic elevators in apartment and office buildings and actually prefer the speed and dependability of service at their convenience rather than at the whim of possibly temperamental attendants.

6.12. *Retail and Wholesale Facilities*

Besides its office buildings and hotels, the central business district attracts people with high-rise buildings of another type, the department store or larger specialty store. Suburban shopping centers are usually planned around one or more such stores.

City or suburban, the store needs a high volume of business through sales to thousands of customers a day coming by mass transportation or family automobile from an extensive, populous area. Success depends to a great extent on "shopper turnover," the total number of persons who, in the course of a day, can reach a store's selling floors, see merchandise on display, and make purchases.

Vertical traffic in such stores is heavy both ways most of the day, a characteristic that has favored escalators ever since their introduction in 1900. Inviting shoppers to reach upper floors without effort, waiting, or crowding, and to view attractive displays en route, escalators continue to be the primary vertical transportation in stores.

Escalators in pairs provide two-way service between principal selling floors, including basement levels (Fig. 6.8). Since safety takes precedence over speed, units operate at 90 fpm and are, of course, reversible. Lower, busier floors usually have escalators 48 in. wide, with units 32 in. wide for the upper floors. In certain parts of the store, traffic may be heavy enough or floors extensive enough to require more than one pair of escalators.

In relation to building plan, escalators should be readily visible and accessible from the main entrances to induce travel to all floors. Upper-floor layouts should be such that escalators stimulate circulation to various selling areas.

6.13. *Department Store Elevators*

Most shoppers, including the many who wish to browse before they buy, use escalators. But some persons, including those whose time is limited, who know exactly what they want, or who suffer physical handicaps, prefer elevators. Store personnel and salesmen

Fig. 6.8. Escalator in "Quelle" department store in Berlin has transparent balustrades to enhance open appearance of selling areas. Helmut Huyde, architect.

calling on the store's buyers also, as a rule, find elevators a more direct means of floor-to-floor travel.

Elevators and escalators in a store are planned to complement one another (Fig. 6.9). As in other commercial buildings, elevators are installed in groups, rather than singly, for frequency and continuity of service. Since selling areas should be visible from all elevator entrances, elevators are laid out in straight-line rather than alcove arrangements, preferably with no more than four cars in a line.

Fig. 6.9. Addition to Charles A. Stevens specialty store, Chicago. Escalators and elevators are designed to complement one another in an integrated system planned for visibility of merchandise as well as rapidity of service.

Because an elevator stops at almost every floor to let a number of shoppers on or off, cars are wider and shallower than in office buildings. Department store elevators are generally of 3,500 lb. capacity, with platforms 8 ft. wide and 5 ft. 6 in. deep (Table 6.2). For wider openings, doors may be of the two-speed center-opening type.

Automatic control of elevator operation and group supervision has found acceptance even in quality stores. Touch buttons and hall lanterns should be prominent, and car position indicators may name the principal department on each floor. Automatic announcing systems may call out floor numbers and special attractions.

In addition to escalators and elevators for passenger transportation, store operation demands freight or service elevators for stock handling. Elevators of 8,000 lb. capacity, with a platform approximately 8 ft. square, large enough to carry even racks of garments, generally suffice for department store service. Rolls of carpet, as long as 15 ft., require an elevator with a sufficiently large platform or high ceiling.

6.14. *Shopping Centers and Merchandise Marts*

While early shopping centers sprawled over extensive areas on a single level, a trend to multilevel, more compact centers, spurred by rising land costs and vanishing sites, is developing. New shopping centers are often planned as part of urban redevelopment projects, where ground space is more limited than in outlying areas. Multilevel design also reduces distances for shoppers between parking fields and the center's stores and services (Fig. 1.5).

Escalators or moving walks (Fig. 6.10) may link the shopping center's various levels. The electrically powered inclined walks carry shoppers in comfort and are especially appreciated by parents with perambulators and persons in wheelchairs. An escalator or moving walk installation can be designed to heighten the center's visual appeal.

Comparable to the department store, except that it attracts wholesale rather then retail buyers, is the merchandise mart. A mart usually specializes in related lines of merchandise, such as apparel or house furnishings, and accommodates "shows" which may each last for a few days. Buyers attending a show attempt to visit all of its sample displays before returning to selected exhibitors to negotiate purchases.

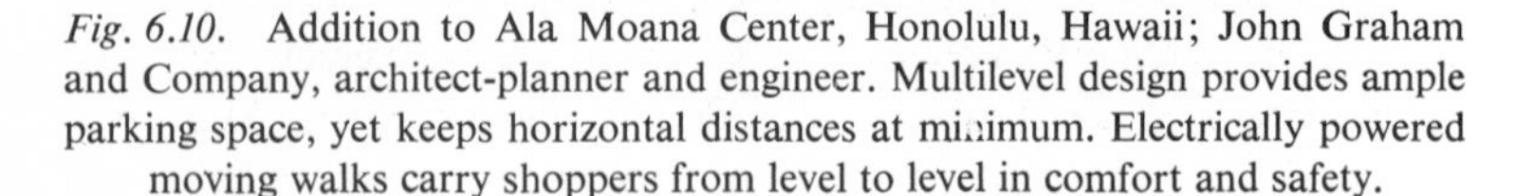

Fig. 6.10. Addition to Ala Moana Center, Honolulu, Hawaii; John Graham and Company, architect-planner and engineer. Multilevel design provides ample parking space, yet keeps horizontal distances at minimum. Electrically powered moving walks carry shoppers from level to level in comfort and safety.

Elevators or escalators carry passenger traffic to the building's upper floors, and there are service or freight elevators for merchandise or displays. Samples rather than merchandise in quantity are usually handled, but setting up and taking down displays in a short time may require considerable elevator capacity.

6.15. *Transportation Facilities*

In airline terminals (Fig. 6.11), bus depots (Fig. 6.12), and railroad and rapid transit stations, people move in steady streams from level to level. Relatively few different levels, moderate total rises, and heavy passenger traffic favor escalators for vertical movement; there are elevators for supplementary service.

Escalators are usually paired for two-way service but can be reversed to handle unusually heavy traffic in one direction or the other. At airports, moving walks may carry passengers over considerable horizontal distances between central buildings and emplaning areas.

In multilevel garages of the ramp type, rising labor costs have contributed to a trend from attendant to customer parking. When a motorist parks his car in a space above or below ground level, he and his passengers require vertical transportation to the street floor.

Because external factors limit the rate at which automobiles can enter or leave a garage, vertical traffic within the facility is unusually moderate in volume and can be readily handled by elevators. Larger, busier public garages, with parallel ramps or lanes to levels above or below ground, also provide escalators for their patrons.

Fig. 6.11. Atlanta, Georgia, Airport Terminal Buildings; Robert and Company Associates, architect. Four escalators, finished in stainless steel, carry passengers between the first and second levels and mezzanine (shown).

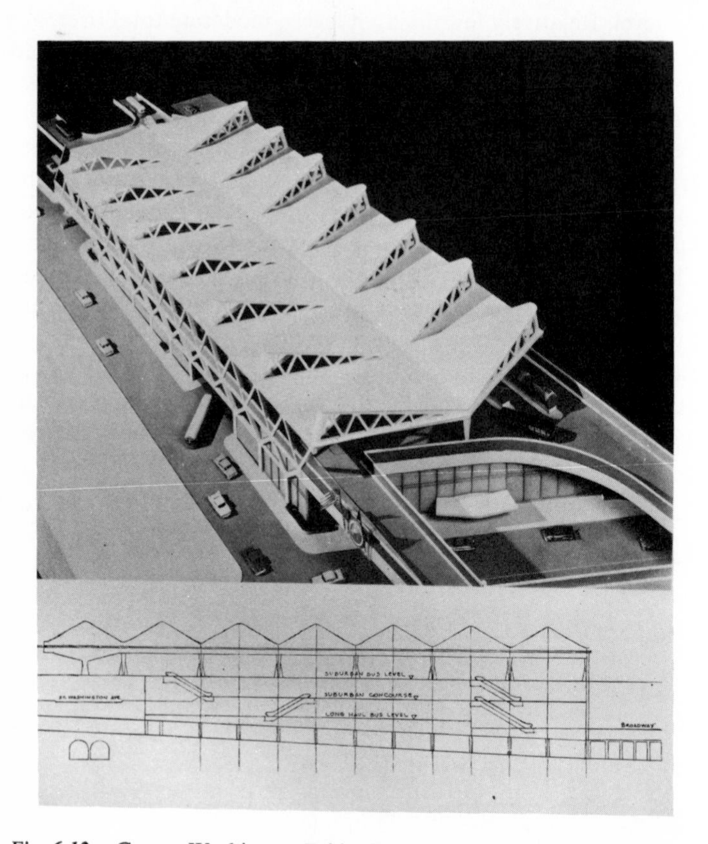

Fig. 6.12. George Washington Bridge Bus Station, New York; Pier Luigi Nervi, architect for roof structure. Entrances from subway and streets are linked to suburban bus platform level by fourteen escalators.

Chapter 7

Residential Buildings

People may require vertical transportation where they live as well as where they work. Some residential buildings today soar to sixty or more stories, as high as many office skyscrapers, but have distinctive transportation requirements because of their distinctive occupancy and use.

This chapter reviews the elevatoring of residential buildings, from two-story private homes to towering apartments. College residences, because their requirements are generally similar to those of apartment buildings, are also treated here, as are the hotels and motels used primarily for residential or recreational rather than commercial purposes.

7.1. *Residence Elevators*

A private home may need an elevator because of its height or to make all of its floors accessible to an ill, elderly, or disabled person who cannot climb stairs. Because it is used by young and old people without special skill, a residence elevator should operate safely and simply, dependably and quietly. Equipment should occupy minimum space and be economical to install in a new or existing home.

Applicable features of modern automatic electric elevators have been incorporated in units designed specifically for homes. Residence elevators (Fig. 7.1) are usually built to carry two passengers, although larger units, for four standing passengers or an attendant with a person in a wheelchair, are available. Traveling at a speed of 35 fpm, these elevators can serve up to four landings with a maximum rise of 35 ft. Simple automatic pushbutton controls facilitate use by any member of a household.

To eliminate need for a penthouse or overhead machine room, the car is underslung and is driven by an electric traction machine in the basement. Car and counterweight run on guide rails forming part

Fig. 7.1. Typical elevator installation in private residence.

of a self-supporting steel frame that relieves the building framework of strain. If the building structure permits, cost may be saved by fastening guide rails and overhead supports directly to the building framework.

Table 7.1
Residence elevators—typical dimensions

Rated passenger capacity	Platform		Hoistway		Doors (standard door height 7′ 0″)
	(w) width	(d) depth	(W) width	(D) depth	clear opening
Two	3′ 0″	3′ 0″	4′ 2″	3′ 11″	2′ 4″
Four, or wheelchair and attendant	3′ 4″	4′ 4″	4′ 6″	5′ 3″	2′ 8″

In an existing residence an elevator may be installed where space is available (Fig. 7.2 and Table 7.1), with minimum building alterations. The hoistway may occupy a stairwell, tier of closets, room corner, part of a hallway, or an exterior shaft (Fig. 7.3).

7.2. *Elevators for Apartment Buildings*

Demanded by rising standards of living and expectations of service, and made feasible by more economical equipment of standardized design, elevators are being increasingly installed even in apartment buildings of only three or four stories. Apartment buildings, from low-rise structures to those towering to office-skyscraper heights where elevators are a prime necessity, share a number of basic requirements for vertical transportation service.

In an elevator apartment building of any height, service should extend to all floors with suites. Skip-stop arrangements, in which an elevator may serve only every second or third floor, have been used in low-rental housing as an economy measure. But the saving of 3 to 5 percent, through elimination of alternate hoistway entrances, is gained at the cost of excessive loss of utility.

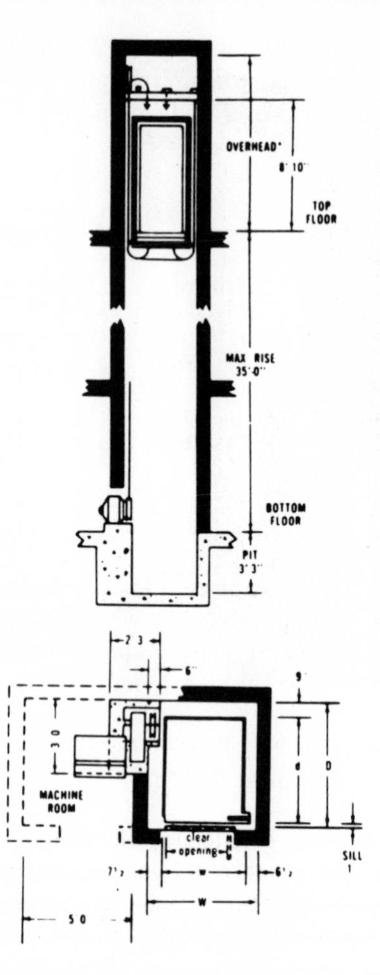

Fig. 7.2. Space requirements for residence elevator. Overhead is approximately 11 ft. 4 in., depending on type of installation and local code. See Table 7.1 for hoistway dimensions.

Fig. 7.3. Residence elevator with hoistway entrance approximately as wide as a closet.

If one car of a pair of skip-stop elevators serves odd floors and the other serves even numbers, average waiting time for passengers on any floor may be excessive. If both elevators of a pair serve the same alternate floors, tenants on the other floors have to use stairs part of the way, a serious inconvenience for handicapped persons and for users of baby carriages and shopping carts.

Levels with tenant facilities such as basement garages and laundry rooms or rooftop sundecks and swimming pools should also be accessible by elevator.

In apartment houses as in other buildings, required elevator capacity depends on the volume of traffic during the busiest period of the day. People come and go throughout the day (Fig. 2.1), but traffic rarely displays the sharp peaks found in office buildings. Traffic increases in the morning when tenants leave for business, in the afternoon when children return from school, and in the evening when tenants return from work. But activity is spread over considerable periods because tenants and their children work and attend school at various distances and at different times.

Evening traffic is usually the heaviest but the number of passengers during any 5-minute period seldom exceeds 6 percent of the building's population. The percentage is higher in a centrally located apartment building occupied largely by working adults than in an apparently similar residence with many occupants above or below working age. Building population may be estimated at 1·5 to 2 persons per bedroom, depending on rental range and other factors.

Requirements for quality as well as quantity of elevator service are also less exacting in apartment than in commercial buildings. While office-building tenants may dislike waiting more than 25 or 30 seconds, the same people, in an apartment, may accept an average wait of 50 to 75 seconds. Apartment-buildings passengers accept somewhat longer riding time than in an office building but balk if the elevator makes too many intermediate stops, or if total trip time—waiting and riding—exceeds 2 or 3 minutes.

Expected elevator service standards vary with rental levels. If upper-floor tenants pay premium rentals, they demand comparably superior elevator service.

Automatic operation, which makes service independent of the availability of labor, has long been accepted in apartment buildings. The old-time elevator attendant, prone to hold an elevator car for some tenants or to leave his car and run errands for others, delayed rather than expedited service. Because unattended automatic elevators provide better vertical transportation at less cost than do attended elevators, it is practical to provide a doorman for security and personal attention.

For passenger convenience and safety modern systems of automatic control include self-leveling. Reducing tripping hazards, accurate automatic leveling is especially valuable in apartment elevators used by persons of all ages and both sexes, possibly carrying babies or bundles or pushing perambulators or shopping carts.

A significant aspect of elevator service, especially in higher-rental apartment buildings, is visual design of entrances and car interiors and quality of illumination and ventilation. In lower-rental projects, however, durability of the installation's exposed parts in withstanding abuse by children may assume priority.

7.3. *Low-Rise and Moderate-Height Apartments*

Many requirements for apartment elevator service apply to buildings of low rise (2 through 5 stories), moderate height (6 through 24 stories), and high rise (more than 24 stories).

In relation to the building plan, elevators should be located centrally, within some 150 ft. of the farthest apartment. If a building has widely separated wings or towers, each unit may require its own elevator or group.

Traffic volume, height of rise, and number of stops determine the size, speed, and number of elevators needed to attain desired service standards. The 1,200-lb. capacity elevator (Table 7.2), which may be used in small three- and four-story apartment houses, has only limited capacity for moving furniture.

Table 7.2
Apartment elevator sizes and capacities

Rated capacity		Platform		Clear hoistway				
lb.	passengers	(w) width	(d) depth	(W) width	(D) depth	Doors, type	Sill	Clear opening (std. ht. 7′ 0″)
1,200	8	5′ 0″	4′ 0″	6′ 4″	5′ 3″	Single-swing	None	2′ 8″
						Two-speed	5½″	3′ 0″
2,000	13	6′ 4″	4′ 5″	7′ 8″	5′ 8″	Single-swing	None	2′ 8″
						Single-slide	4″	3′ 0″
						Center-opening	4″	3′ 0″
2,500	16	7′ 0″	5′ 0″	8′ 4″	6′ 3″	Center-opening	4″	3′ 6″

Passenger elevators for larger buildings should have 2,000 lb. lifting capacities and car interiors at least 6 ft. wide and 3 ft. 8 in. deep. This size can comfortably hold eight to ten passengers or a perambulator and two or three standees. The car's wide, shallow shape lets passengers on and off more rapidly at each stop, improving service. An elevator of this size and shape also facilitates moving.

For a building up to 6 stories high with not more than 75 apartments, one such elevator is normally adequate. Taller buildings need two or more cars operating as a group. At least one elevator is then always available.

A separate service elevator, located near the building's delivery entrance, should have a minimum capacity of 2,500 lb. and an interior 6 ft. 8 in. wide by 4 ft. 3 in. deep. A car this size, with high ceiling and wide doors, accommodates furniture up to 8 ft. long. If one of the passenger elevators is to be used for service purposes part of the time, it should have removable wall pads to protect the finish of the car when furniture or freight is being handled.

Elevator speed depends primarily on the height of the building. For installations up to six stories, speeds up to 150 fpm are common. For six to ten stories, speeds range up to 200 fpm and for ten or more stories, to 350 fpm or more. Machines may be hydraulic (Fig. 4.4) or geared (Fig. 7.4 and Table 7.3) for rises up to five stories, geared for six to ten, and gearless for taller buildings (Fig. 7.5).

Table 7.3
Typical pit depths and overhead heights

Speed (fpm)	Capacity (lb.) 1,200		2,000		2,500	
	pit	o. ht.	pit	o. ht.	pit	o. ht.
100–150	4′ 0″	14′ 5″	5′ 0″	14′ 10″	5′ 0″	14′ 10″
200	—	—	5′ 0″	14′ 8″	5′ 0″	14′ 8″
250–300	—	—	5′ 0″	15′ 2″	5′ 2″	16′ 2″
350	—	—	5′ 0″	15′ 2″	5′ 2″	16′ 2″
500	—	—	8′ 2″	17′ 4″	8′ 2″	17′ 10″

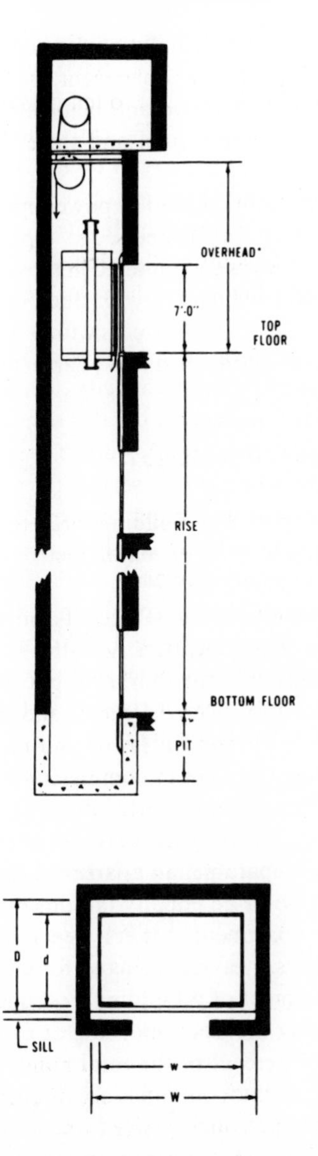

Fig. 7.4. Space requirements for apartment elevator. See Table 7.2 for hoistway dimensions, and Table 7.3 for pit depths and overhead heights, which may vary with local codes.

Automatic operation is usually of the collective type with provision, particularly in luxury apartment buildings, for optional attendant service. A group of two automatic elevators may be supervised by a duplex collective system and three or more, by multiple zoning.

Car and hoistway doors should be power operated with automatic door-reversal devices of the rubber-edge type providing sufficient protection under conditions encountered in most apartment buildings. Elevators up to 2,000-lb. capacity with door openings to 3 ft. have single-slide entrances, and 2,500-lb. cars with 3 ft. 6 in. door openings, center-opening or two-speed entrances (Table 7.2).

7.4. *High-Rise Apartment Buildings*

Construction of apartment buildings of 25 to 30 stories and higher has been increasing, not only in major metropolitan centers but in smaller cities as well.

Rising costs of land have encouraged its more intensive use, often to make the most of a choice location. As with offices, apartment rental scales depend not only on neighborhood but also on height. Such high-rise advantages as privacy, a view, and natural daylight, in addition to the release of land for lawns and play areas, may weigh more heavily in residential than in commercial buildings.

Desirability of upper-floor apartments also depends on effective vertical transportation under a distinctive set of conditions. Although the heights may be comparable, an apartment building's needs differ significantly from those of an equally tall office building.

Since in-town apartment towers generally house employed individuals and couples, heaviest demand for elevator service occurs in the morning when most of a building's occupants leave for work. Every 5 minutes during the peak, morning or evening, elevators may have to handle 5 to 7 percent of the total building population.

Tenants of higher-rental buildings are frequent users of part-time maid, valet, and other personal services which generate considerable service-elevator traffic. Since service traffic slackens in the early evening, before dinner, a properly located service elevator can supplement the passenger cars during this period of heavy passenger traffic.

Fig. 7.5. Edificio Gualanday in Medellin, Colombia, has 38 luxury apartments on floors 2 through 24, professional offices on the ground floor, and parking facilities at the basement level. Two gearless elevators provide vertical transportation. Arcila Wills Cordoba y Cia. Ltda, and Ingenieria y Construcciones Ltda, architects; owners, Propiedad Horizontal Gualanday.

Service elevators may be even more essential in high-rise apartment buildings than in office skyscrapers. Rapid turnover of downtown apartment tenants produces moving-day traffic which ties up an elevator for prolonged periods, often including the morning peak.

Higher rises not only demand faster elevators for transportation of sufficient quality and quantity (Table 7.4), but the longer runs also let elevators travel at their top speed more of the time. Passenger elevators are usually in the speed range for gearless machines, and cars are grouped, often for local and express service.

Table 7.4
Typical elevator travel speeds for high-rise apartment buildings

	Speeds (fpm)	
Stories	Passenger	Service
10–20	200–350	200
20–30	350–500	350
30–40	500–700	500
40–50	700–1,000	500
50–60	1,000–1,200	700
More than 60	1,200 and over	800

Plans for elevatoring a 34-story apartment building with a population of approximately 1,410 suggest the excellent service standards that may be attained with local and express operation (Fig. 7.6).

Local elevator speeds of 350 fpm enable a three-car group to carry 53 persons in a 5-minute period, or 6 percent of the population of 870 on the floors served, with a 61-second average interval between elevators. The three express elevators, traveling at 500 fpm, have a 5-minute handling capacity of 38 persons, or 7 percent of the population of 540 on the upper floors, with a 62-second interval.

Studies for a 30-story building with a smaller population indicated that three elevators, serving all floors, would provide adequate carrying capacity and an acceptable interval. But four elevators, two local for floors 1 to 15 and two express for floors from 16 up, were found to provide better service without substantially increasing elevator costs.

Instead of three high-speed elevators, the express-local system would need only two high-speed and two slower elevators, with 60 entrances instead of 90 for all-local operation. Total space requirements for hoistways would be the same with either arrangement (Fig. 7.7). The local-express plan, however, offers the added gain of

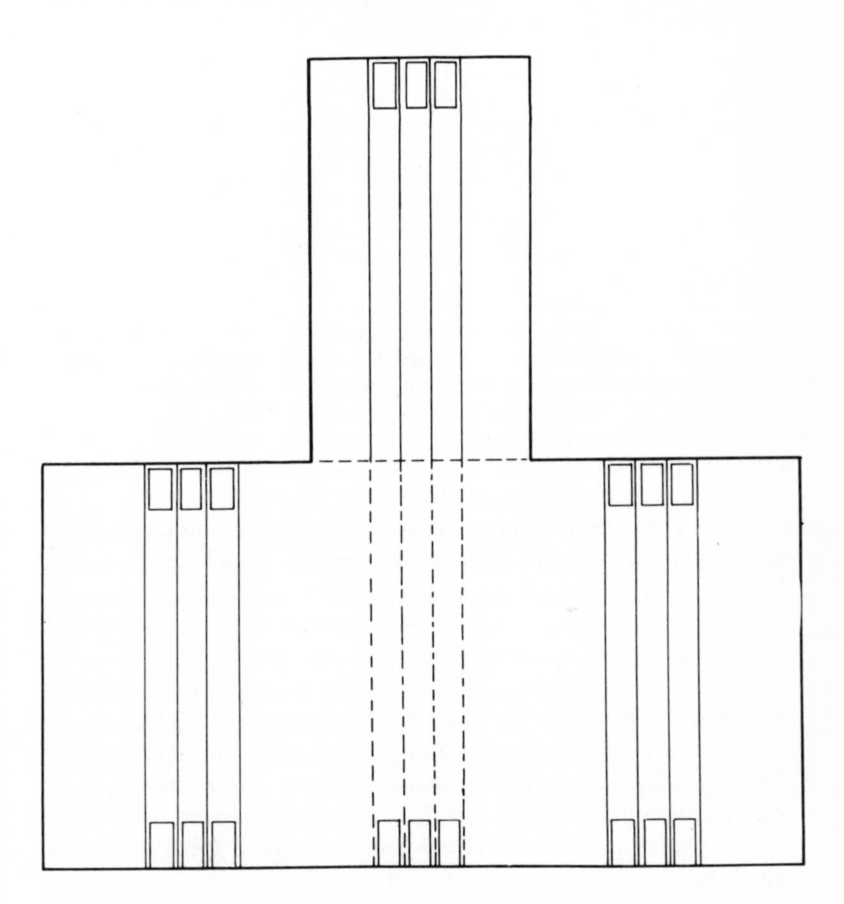

Fig. 7.6. High-rise apartment building with express elevator group in tower, local groups in wings.

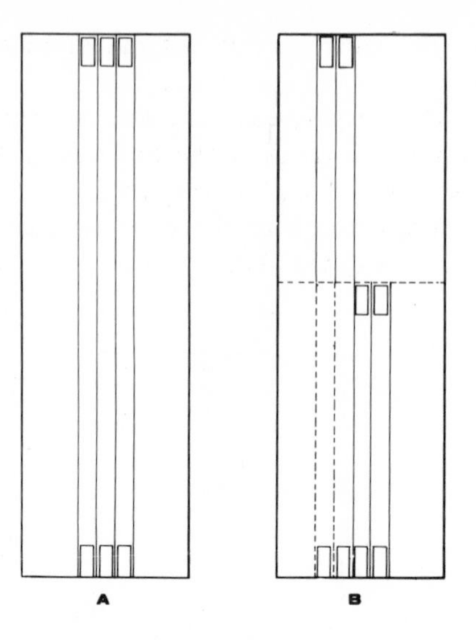

Fig. 7.7. Four elevators—two express, two local (A)—may offer better service at about the cost of three local elevators (B). Express-local plan needs less hoistway construction at upper floors, one-third fewer hoistway entrances.

reduced space consumption on upper floors where construction costs more and rentable space can earn greater returns.

For prompt, automatic response to passenger calls, each group of elevators in the high-rise apartment building should be controlled by multiple zoning, with necessary circuits for handling periods of peak traffic. Especially if such traffic is expected, door reversal devices of the electronic detector type may be advisable.

Office building-type signals find increasing use in apartment skyscrapers, especially at main-floor entrances and for larger groups of elevators. Directional lanterns alert a passenger to an elevator going his way, while in-car position indicators show him plainly when he reaches his floor. Besides aiding the individual passenger, such signals encourage faster loading and unloading of cars to speed service for all users.

7.5. *Housing for the Elderly*

In requirements for vertical transportation, housing for the elderly resembles conventional apartment buildings. One difference is the greater dependence of older persons on elevators even in buildings of only two stories. Needs are also modified if a building includes facilities like those in nursing homes and has single suites as well as housekeeping apartments. But the distinguishing characteristic of elevator service in residences for the elderly is the emphasis on safety and comfort rather than speed and capacity.

In addition to the usual safety provisions, elevators should have accurate automatic leveling, which is of particular value in reducing tripping hazards at entrances and facilitating movement by passengers in wheelchairs. Light beam as well as rubber safety-edge systems should be installed to protect users against premature door closing. For passenger comfort, acceleration and deceleration should have the smoothness afforded by variable voltage control. Cars should be well lighted and fitted with handrails and, preferably, a telephone for emergencies.

In typical housing for the elderly, service demand is seldom concentrated at particular times unless some of the residents use communal dining on a central floor. Elevators are usually automatic, with controls of the collective type and power-operated doors. Car and landing buttons should be low enough to be easily reached from a wheelchair, and all indications should be large and plainly legible. Passengers are more tolerant of waiting but, because they react more slowly, automatic control circuits should allow doors to remain open longer at stops.

In addition to one or more passenger elevators, usually of the 2,000-lb. apartment type, a hospital-shaped service elevator (Section 8.2) of 3,500 lb. capacity should be available in the larger building for stretchers and beds. The service elevator may have a rear entrance near a building driveway for ambulances (Fig. 7.8). In a smaller building, a 2,500-lb. elevator with two-speed doors opening to 3 ft. 6 in. can be arranged to accommodate a portable stretcher.

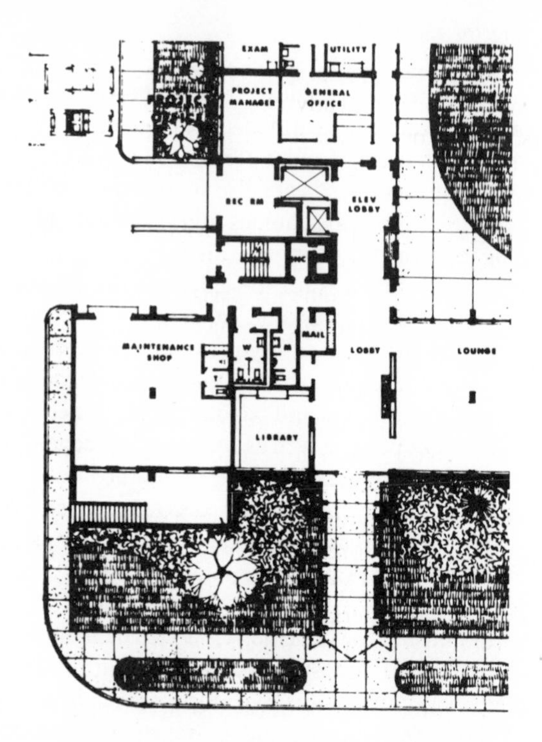

Fig. 7.8. Ground-floor plan for housing for elderly, showing, in addition to passenger elevator, hospital-shape service car with rear entrance to ambulance driveway. Warren (Ohio) Metropolitan Housing Authority, Riverview Housing for the Elderly; Arthur F. Sidells, architect.

7.6. *College Residences*

Concurrent problems of housing rapidly expanding student populations, limited funds for land acquisition, and higher land prices are forcing more colleges and universities to grow upward rather than outward. High-rise residence buildings, long a feature of metropolitan schools, are becoming increasingly common on outlying campuses.

Land cost and availability are not the only forces behind the trend to taller structures. As an institution grows, spreading academic and residential buildings create problems of distance, time, and transportation which invite solution through multistory design. Tall buildings also permit grouping facilities floor by floor, with administrative and psychological advantages over strung-out plans with long corridors. Dormitories stacked vertically rather than extended horizontally yield appreciable savings in the cost of heating, cooling, and other services.

When residence buildings rise beyond two or three stories, elevator service assumes major importance. Fully realizing the advantages inherent in high-rise design demands efficient vertical transportation for on-time movement of students to and from other campus facilities.

Most elevator traffic moves directly between the main and upper floors, as students go to and from their rooms to classes, meals, and other activities. A rooftop lounge or other upper-floor attraction influences the traffic pattern.

In a dormitory the traffic peak which usually determines elevator capacity occurs just before the evening meal, if the dining room has conventional waiter service and students must be seated at table at a stated time. With cafeteria-style feeding, elevator traffic reaches its peak about the middle of the evening meal.

Traffic during this period depends largely on the total population of the building, which may be known or may be approximated on the basis of 100 to 120 sq. ft. of usable floor space for every student-resident. Surveys in existing residences show that, during the dinner hour peak, 10 to 15 percent of a building's population may demand elevator service every 5 minutes. This percentage may be less if limited, for example, by cafeteria serving capacity.

Size, speed, and number of elevators depend on the number of people to be carried and number of floors to be served. In addition to providing the quantity of transportation required to handle traffic loads, a residence building installation must provide service prompt enough that passengers need not wait more than 75 seconds for an elevator. Skip-stop arrangements are usually undesirable, for reasons similar to those applying to apartment buildings.

Since adequacy of elevator service depends on capacity and performance of elevator equipment in relation to traffic load, planning may well begin with measures to reduce demands on the building's elevators. Traffic can be minimized by locating heavily used facilities, such as restaurants or central recreation rooms, on or near the ground floor.

A residence of six stories or more requires two or more elevators located side by side and operated as a coordinated group. Automatic control systems are like those in apartment buildings, with special provisions to resist the greater exuberance of students.

7.7. *Residential Hotels and Motels*

Some hotels and motels, as has been suggested, are characterized by a rapid pace of activity and require vertical transportation comparable in capacity and frequency to other commercial buildings. But in other hotels and motels (Fig. 7.9) the pace and pattern of vertical

Fig. 7.9. Motor inn on sloping site in Kansas City, Missouri. Five floors of guest rooms are linked to main floor by two elevators that also give guests direct access to parking areas and swimming pool without going through lobby.

traffic more closely resemble that of the apartment than the office building.

Apartment hotels may, for elevator planning purposes, be classed in the residential rather than the commercial category. So may hotels and motels patronized by individuals and families primarily on vacation rather than on business trips, especially if they travel by private automobile rather than air, road, or rail mass transportation.

Hotels and motels in the residential category seldom have convention facilities or room service to intensify demands for vertical transportation. But all floors used by hotel or motel guests, including basement parking and rooftop recreation and refreshment facilities, need appropriate elevator or, in some cases, escalator service.

Elevator capacity must be sufficient to handle traffic during the day's two busier periods, in the evening as guests check in and in the morning as they depart. At these times, traffic every 5 minutes may be as much as 10 percent of the guest population.

If the building has an upper-floor restaurant, its traffic, which depends on seating capacity and turnover rate, is added to the elevator traffic to guest-room floors. For most hotels and motels in this category, elevator service frequency from 40 to 60 seconds is satisfactory.

In the absence of rooftop dining facilities, separate service elevators are seldom necessary. Motel guests usually carry their own luggage, chambermaids can use passenger elevators between periods of busiest traffic, and room service is minimal. A rooftop restaurant, however, may require a service elevator to deliver food and beverages and to remove waste.

For buildings up to four stories, elevators may be spotted between parking areas and guest rooms to minimize walking distances. In taller inns, elevators should be grouped, near the main entrance, to minimize waiting time.

Passenger elevators are usually of 2,500 or 3,000 lb. capacity, with cars 7 ft. wide by 5 ft. or 5 ft. 6 in. deep. The smaller car comfortably accommodates 10 to 12 passengers without luggage or 6 persons with one large suitcase each, while the larger car readily takes 14 to 16 passengers. Center-opening doors with clear entrances 7 ft. high by at least 3 ft. 6 in. wide easily admit a person carrying a suitcase.

Service elevators, where used, are the 3,000- or 3,500-lb. size. The latter has a platform 7 ft. by 5 ft. 6 in. and can readily carry hand trucks and food carts to rooftop facilities.

Elevators operate automatically, with control systems like those in apartment buildings. Grouped passenger elevators are coordinated by duplex collective or multiple zoning systems, with circuits for continuous service during periods of peak traffic.

Chapter 8

Institutional Buildings

Multistory buildings, with concomitant requirements for vertical transportation, have long been the rule for hospitals. In recent years more effective utilization of available personnel has heightened interest in buildings of tall, slender design. In other types of institutional buildings for schools, libraries, and museums the quest for better site planning as well as improved operation often leads to high-rise construction and increased emphasis on vertical transportation.

8.1. *Vertical Traffic in Hospitals*

Whether located in the heart of a city or in an outlying area, modern hospitals are usually housed in multistory buildings to conserve not only ground space but also the time and energy of staff members. With high-rise design, short, swift elevator or escalator rides can eliminate long, time-consuming, and tiring walks.

Vertical transportation for a hospital, as for other types of buildings, should be related to anticipated volume and distribution of vertical traffic as well as to building plan.

Traffic is usually considerable during the morning and early afternoon; then there is heavy passenger traffic during evening visiting hours (Fig. 8.1). In any particular hospital, traffic patterns throughout the day are influenced by schedules for treatment, feeding, visiting, and other activities. Volume during busy periods, particularly when nurses and other staff members are changing shifts, imposes critical demands on vertical transportation capacity.

Maximum traffic intensity depends on a hospital's population, primarily staff and students. During a critical 5-minute period, elevators may be called on to carry some 10 to 12 percent of the total number of people in these categories plus essential vehicle traffic. Total passenger traffic consists of staff, volunteers, students (in a teaching hospital), visitors, and, to a limited extent, patients.

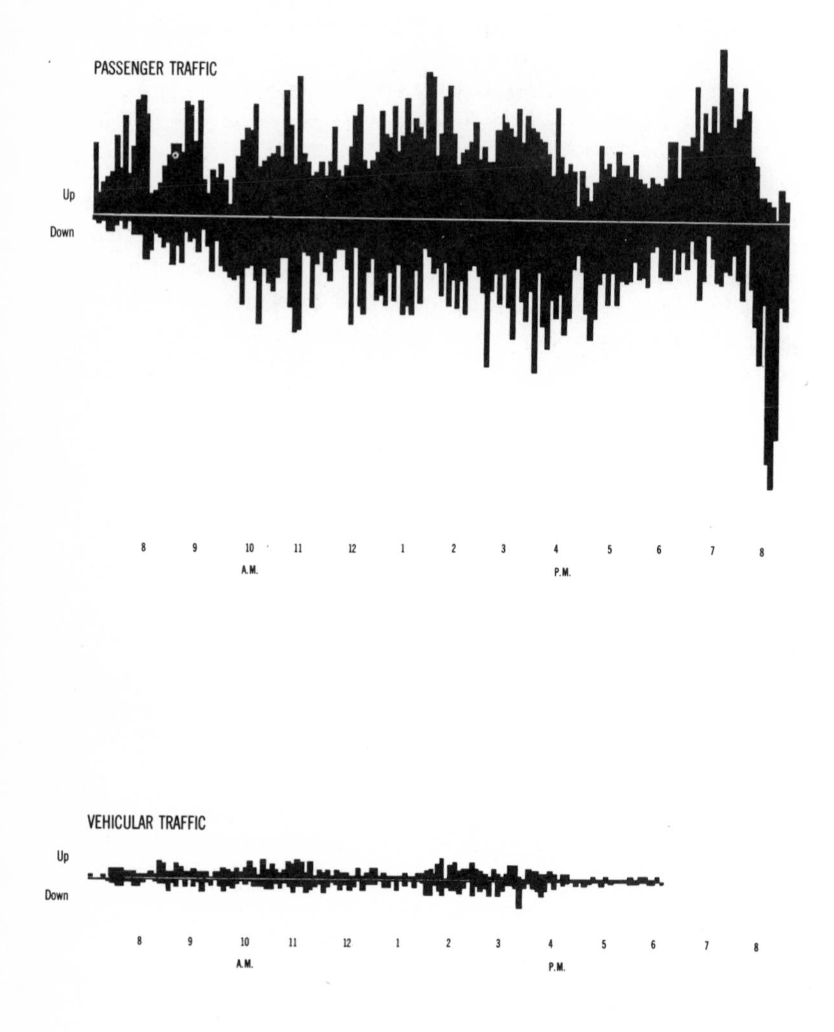

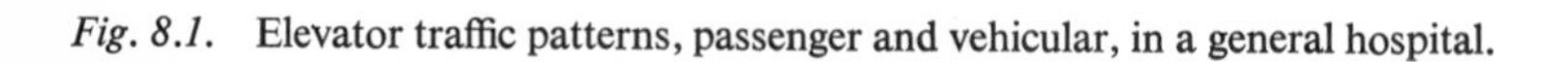

Fig. 8.1. Elevator traffic patterns, passenger and vehicular, in a general hospital.

Staff-to-patient ratios have been rising with shorter employee working hours and more frequent shift changes, more rapid patient turnover, and more intensive treatment. Semiprofessional and technical workers account for a growing proportion of all personnel.

Staff traffic consequently imposes increasingly greater demands on hospital elevators. Visitor traffic, on the other hand, is less critical than formerly, as visiting hours have become more extended and elevators less subject to concentrated peaks.

Hospitals, with at least some staff on duty during all hours of the day and night, seldom experience the morning incoming and afternoon outgoing peaks of office buildings. But hospital elevators must have sufficient capacity for periods of heavy staff traffic, when new shifts of nurses go on duty, doctors visit patients, surgical cases are moved to operating rooms, cleaning and maintenance people make their rounds, and serving the noon meal begins. In teaching hospitals, elevators also carry groups of students visiting their assigned patients, or classes of nurses moving through the building.

Traffic is heaviest to and from floors with central supply, laboratory, X-ray, or cafeteria facilities. An operating-room floor without intensive care units is also a source of concentrated elevator traffic. If outpatients must travel to upper floors, they add appreciably to necessary elevator capacity.

With greater proportions of skilled people on hospital staffs and continually rising salaries and fringe benefits, it becomes increasingly urgent to reduce elevator waiting time to a minimum. About 40 seconds may be acceptable in smaller hospitals, but 30 seconds is preferable for larger institutions.

Even a hospital with only two stories needs a pair of elevators, so that at least one continues in operation while the other is being serviced. Installing and operating elevators in groups rather than singly reduces waiting time.

Elevator travel speeds, which should increase with the height of a hospital (Table 8.1), influence the quantity and quality of service. Gearless elevators, with their high speeds (350 fpm and more) and smooth acceleration, provide the performance required for hospitals taller than six stories. For hospitals of fewer floors, geared-type elevators of 350 fpm or less economically satisfy service requirements.

Table 8.1
Recommended elevator speeds for hospitals

2–3 floors	100–200 fpm
3–5 floors	200–350 fpm
6–8 floors	350–500 fpm
8–12 floors	500–700 fpm
Over 12 floors	600 fpm or higher

Riding time may be minimized by control systems that save time in leveling and door operation and make elevators accelerate and decelerate at the most rapid comfortable rates. If the number of intermediate stops is reduced by avoiding front and rear entrances for the same elevator and, if feasible, by separating elevators into local and express groups, users also enjoy more prompt service.

In the interest of service quantity as well as quality, elevators should have only one lower terminal, and that on the main entrance floor. Stops at multiple-entrance levels consume too much time, decreasing total service unless more or faster elevators are used.

If the hospital site requires entrances at more than one level, entrance floors may be linked by shuttle elevators or escalators. Such an arrangement often achieves better service than do high-rise elevators stopping at all entrance floors.

8.2. *Passenger and Vehicle Traffic*

People account for most of the elevator traffic, but an important element (Fig. 8.1, lower graph) consists of vehicles: stretchers and wheelchairs, portable apparatus, and food and supply carts. Volume during a busy 5-minute period may be as great as 3 to 4 vehicles per 100 beds.

In a building with two or three elevators, all should be of the hospital service type, deep cars designed to accommodate vehicles and their attendants (Fig. 8.2, Tables 8.2 and 8.3). The smaller of these elevators have a capacity of 4,000 lb., but larger cars of 4,500 lb. or more permit easier handling of beds or bulky mobile equipment.

In larger hospitals, with four or more elevators, separate groups for vehicular and passenger traffic have proved advantageous.

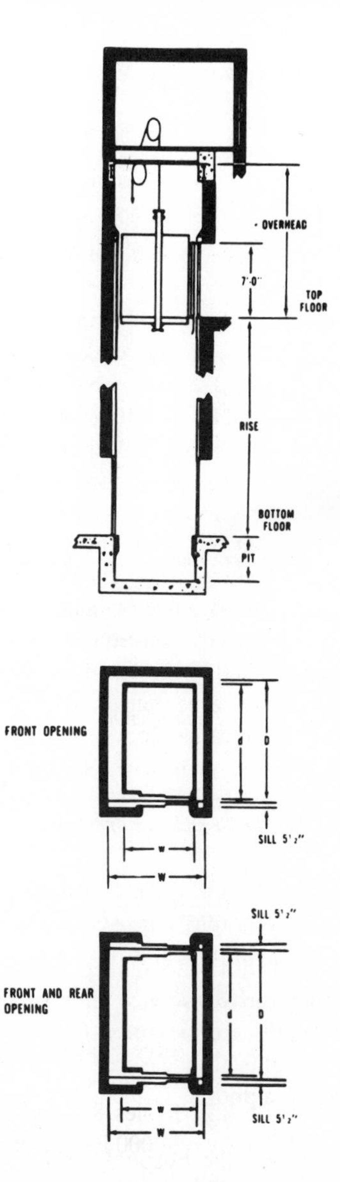

Fig. 8.2. Simplified elevation and plans for hospital elevators. Typical dimensions appear in Tables 8.2 and 8.3.

Table 8.2
Hospital elevator capacities and dimensions

Rated capacity		Platform		Clear hoistway		Doors (standard door height 7′ 0″)
pounds	passengers	(w) width	(d) depth	(W) width	(D) depth	clear opening
4,000						
Front opening	27	5′ 8″	8′ 8″	7′ 8″	9′ 1″	4′ 0″
Front and rear opening	27	5′ 8″	9′ 1½″	7′ 8″	9′ 6″	4′ 0″
4,500						
Front opening	30	5′ 10″	9′ 0″	7′ 11″	9′ 5″	4′ 0″
Front and rear opening	30	5′ 10″	9′ 6″	7′ 11″	9′ 10½″	4′ 0″

Table 8.3
Typical pit depths and overhead heights for hospital elevators

75 fpm		100 fpm		150–200 fpm		250 fpm		400 fpm		500 fpm	
pit	o. ht.	pit	o. ht.	pit	o. ht.	pit	o. ht.	pit	o. ht.	pit	o. ht.
Front opening											
5′ 0″	15′ 0″	5′ 0″	15′ 0″	5′ 0″	15′ 0″	6′ 9″	15′ 4″	8′ 8″	22′ 0″	9′ 2″	24′ 0″
Front and rear opening											
5′ 0″	16′ 8″	5′ 0″	16′ 8″	5′ 0″	15′ 0″	6′ 9″	16′ 4″	8′ 8″	25′ 0″	9′ 2″	27′ 6″

Separating the two kinds of traffic to keep them from interfering with each other markedly improves service all around.

Minimum pit depths are measured from lower landing floor to pit floor. Overhead heights are measured from upper landing floor to top of machine beam supports. Dimensions may vary slightly with the local code.

Vehicle elevators are the hospital service type. But pedestrian traffic is carried on a group of passenger elevators, with wide, shallow cars and center-opening doors to let people in and out quickly.

Recently built hospitals have used such elevators of 3,500 lb. capacity, large enough for as many as 23 passengers or an ambulance-type stretcher in an emergency.

Increasingly, hospitals are using escalators for heavy, continuous pedestrian traffic to outpatient departments up one or two stories. Visitors and personnel may also ride escalators from a street-level entrance to a main lobby higher. Diverting much of this traffic from the elevators enables them to provide improved service to the upper, nursing floors.

8.3. *Hospital Elevator Layout*

Elevators form a vertical service core in the hospital building, often at the intersection of the wings in a building with a Y or X floor plan or at the center of a circular hospital. In one plan (Fig. 8.3) elevators are in a utility tower between two circular nursing wings. To shorten the most heavily traveled paths of horizontal movement, elevators should be located near the hospital's center of population and as close as possible to intensively used facilities.

Passenger elevators should open on a ground-floor passenger lobby accessible to the main building entrance, service cars on separate service corridors accessible to the emergency entrance and the loading dock (Fig. 8.4). If a hospital has only three elevators, they may nevertheless be planned to provide, in effect, two groups—passenger and service—of two cars each.

In the latter case the three elevators are located between passenger and service corridors, the middle car having front and rear entrances opening on both corridors (Fig. 8.5). Since peaks of vehicular and passenger traffic are seldom concurrent, this arrangement makes two cars available to handle peak loads of either type. Assignment of the center car to either the passenger or the service corridor can be controlled automatically, by clock or keyswitch. Planning the elevator core should allow for future expansion, horizontal or vertical, of the hospital building, with at least one extra hoistway (Fig. 8.6), depending on the ultimate plan. Traction-type driving machines located below the lower landing, rather than in a penthouse atop the hoistway, need not be raised if elevator service is later extended to serve added upper floors.

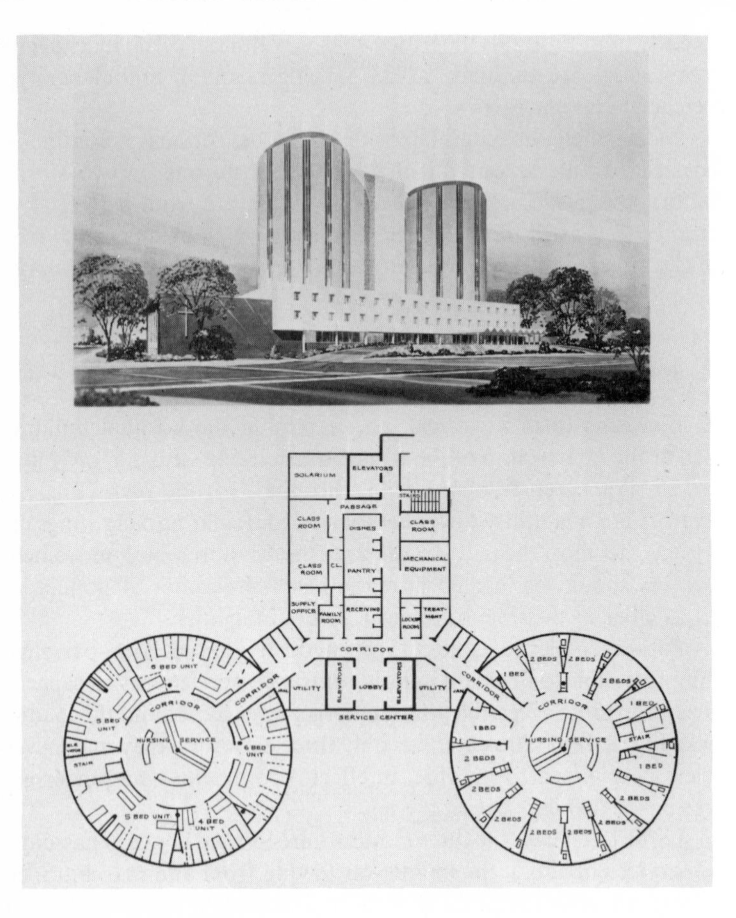

Fig. 8.3. St. Joseph's Hospital, Denver, Colorado, combines advantages of round and rectangular design. Nurses' stations in circular towers are close to patient rooms around perimeter and to elevators and service facilities in rectangular central tower.

8.4. *Automatic Control Systems*

For reasons of performance as well as economy, automatic control has long predominated in hospitals. For a two-elevator group in a low-rise hospital, operation is usually duplex collective, multiple zoning being preferred for taller, busier hospitals with groups of three or more cars.

Special controls may be incorporated in a system to permit authorized persons to detach one car from a group temporarily, should priority movement of a patient from one floor to another be required.

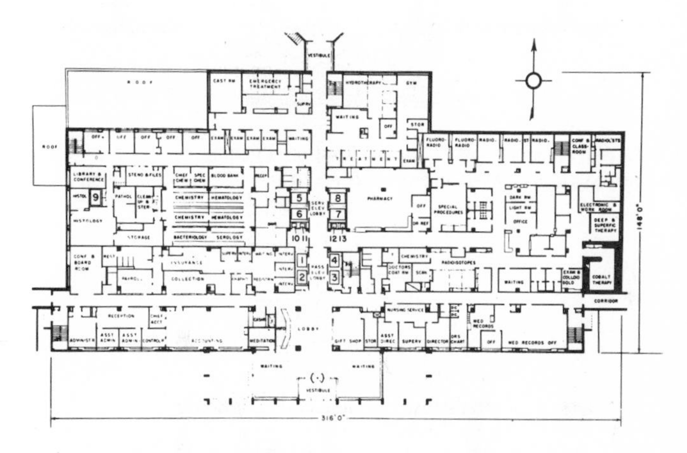

Fig. 8.4. Ground-floor plan of Riverside Methodist Hospital in Columbus, Ohio. Elevators 1–4 are passenger; 5–8, service; 9, freight. Dumbwaiters are 10–13. Schmidt, Garden and Erikson, architect; Inscho, Brand & Inscho, associate architect and engineer; G. W. Atkinson and Son of Columbus, general contractor; owner, White Cross Hospital Association of Ohio.

In a well-elevatored hospital up to five floors high, the probable number of intermediate stops is usually so few that an average elevator trip seldom exceeds 60 or 80 seconds. Under such circumstances special controls for nonstop operation are unnecessary and would only impair service for the hospital as a whole.

In a taller hospital the greater number of probable intermediate stops may warrant provisions for priority service. For this purpose a keyswitch may be installed to cause the elevator to bypass hall calls.

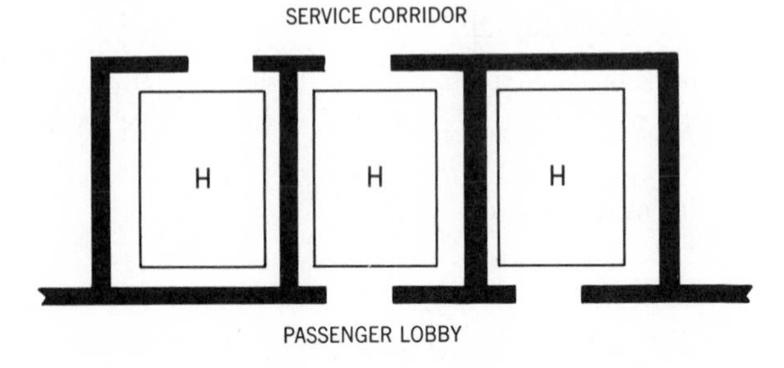

Fig. 8.5. Arrangement with three elevators opening on hospital service and passenger corridors. Assignment of center car must be controlled by hospital administration.

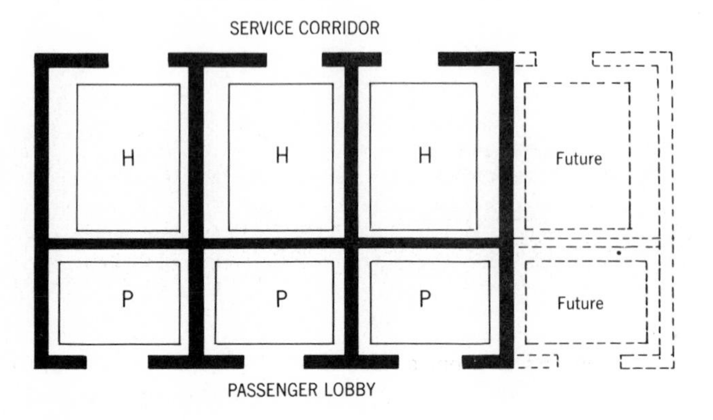

Fig. 8.6. Grouping hospital service and passenger elevators with provision for future hoistways.

8.5. *Food Service and Supplies*

Although elevators are more flexible, automatically controlled electric dumbwaiters may be more economical for certain functions. Dumbwaiters also conserve labor, since they do not require attendants to ride with food or supply carts from floor to floor. Handling carts on dumbwaiters can relieve traffic on elevators during critical periods and thereby improve service for other users.

Recent developments in dumbwaiters enhance their applicability to hospital transportation.

Floor-level loading dumbwaiters, for instance, save time and effort in food distribution (Fig. 8.7). Large enough to carry an entire

Fig. 8.7. Floor-level dumbwaiters in hospital carry food carts between central kitchen and nursing floors.

tray cart, these dumbwaiters eliminate loading and unloading of individual trays required by conventional dumbwaiters. Counter- or floor-level dumbwaiters with ejector equipment automatically load or unload the car at desired floors.

Ejector dumbwaiters (Fig. 8.8) avoid distracting a nurse from other

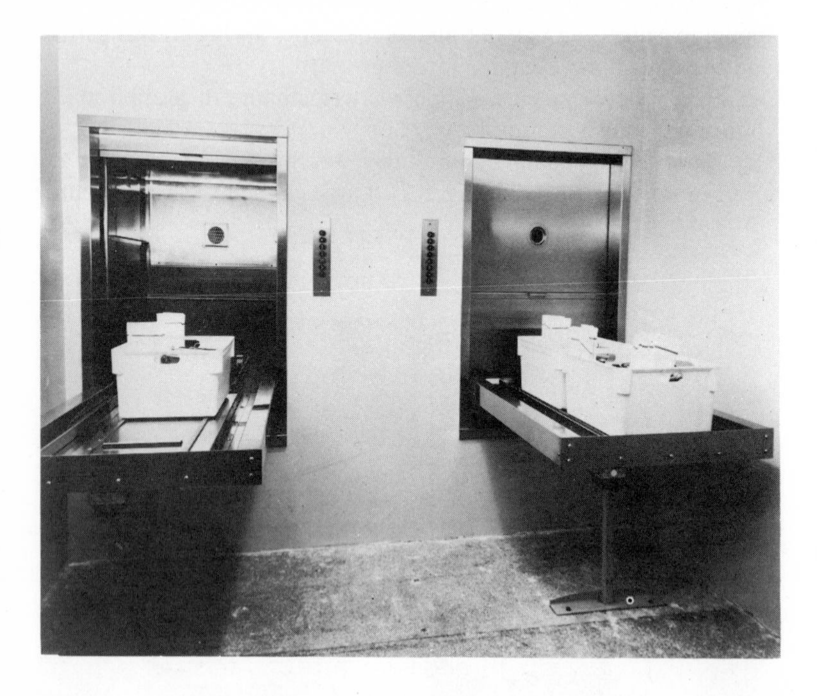

Fig. 8.8. Ejector dumbwaiters relieve demands on hospital personnel by automatically unloading supplies at designated floors and returning to dispatching station. This form of automation may be applied to floor-level as well as counter-height dumbwaiters.

tasks to take pharmaceuticals or supplies off the car each time it arrives. Instead, the dumbwaiter doors open automatically, the mechanism in the car ejects its load, and the car then returns to the floor from which it was dispatched.

Full automation is achieved with dumbwaiter-type lifts that pick up as well as eject carts at desired floors, so that an attendant no longer need be present to load the lift (Fig. 8.9). He simply places the hospital cart in front of the lift door, presses a button, and goes about other duties. When the lift arrives, its doors open and a conveyor mechanism inside the car extends and engages the underside of the cart. The mechanism retracts, drawing the cart into the lift, the doors close, and the cycle of automatic operation continues as with the ejector dumbwaiter.

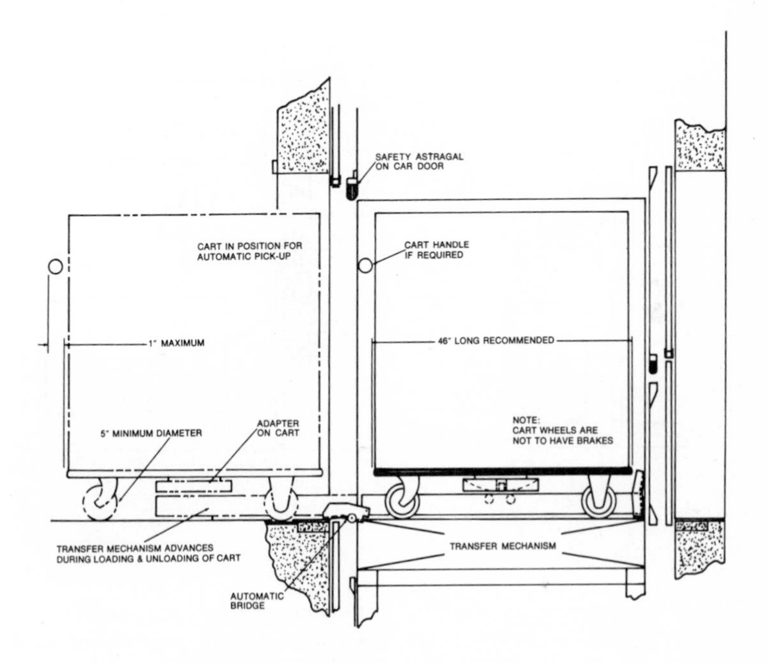

Fig. 8.9. Dumbwaiter-type lift with automatic mechanism for loading as well as ejecting hospital food and supply carts.

Similar arrangements automate the handling of meal trays and supply containers. Developments like these make modern dumbwaiters a highly flexible form of automatic vertical distribution, since they can be installed in hoistways of various sizes and their travel can be extended if more floors are added to a hospital.

Overhead conveyor systems that automate horizontal movement

of food and supply carts can be extended vertically to many floors of a high-rise hospital with special lifts (Fig. 8.10). Integrating horizontal and interfloor cart movement under automatic control, the new system speeds deliveries between a central facility and locations throughout the hospital.

Operation is fully automatic, allowing personnel to remain at their regular stations and continue normal duties. Control system

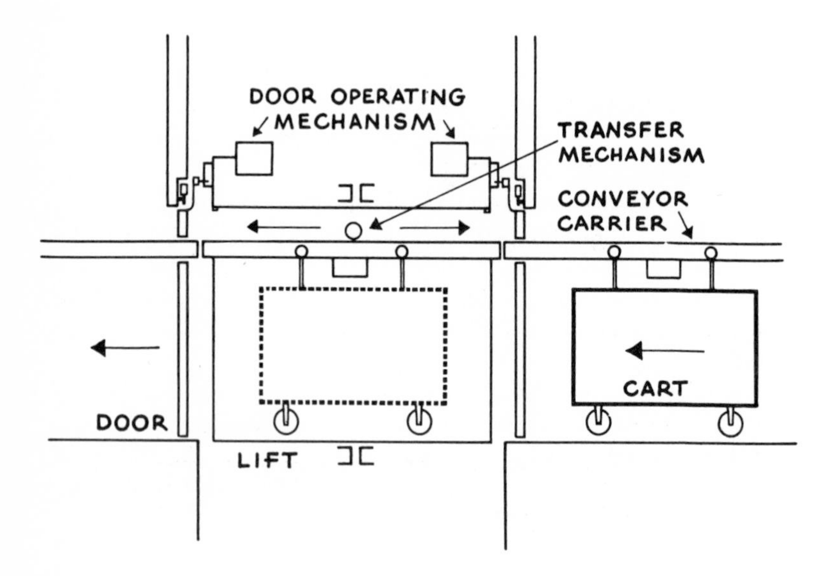

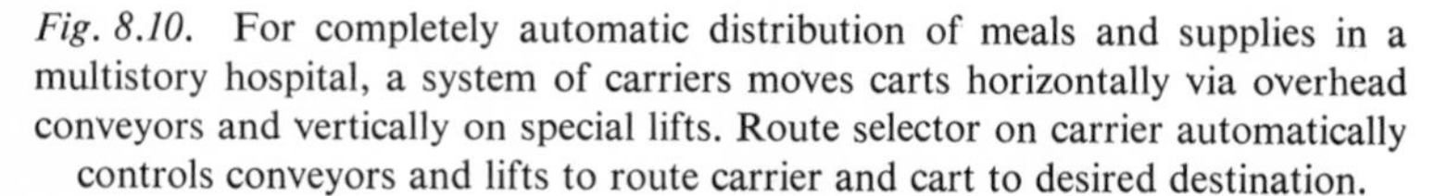

Fig. 8.10. For completely automatic distribution of meals and supplies in a multistory hospital, a system of carriers moves carts horizontally via overhead conveyors and vertically on special lifts. Route selector on carrier automatically controls conveyors and lifts to route carrier and cart to desired destination.

flexibility simplifies dispatching carts in either direction between various origins and destinations.

Hospital carts are suspended off the floor from carriers moved horizontally by overhead conveyors and floor-to-floor by automatic lifts. Each carrier has a route selector which can be set for the desired destination. Contacts on the selector close circuits that automatically call the lift and send it to the proper floor and switch the carrier to the proper conveyor lines.

Lift doors remain closed unless a carrier and its cart are entering or leaving. When a carrier and cart reach the lift hoistway and the lift arrives, its doors open and an automatic transfer mechanism draws the carrier into the lift. At the destination floor the mechanism ejects carrier and cart to the overhead conveyor for horizontal movement.

8.6. *Convalescent and Nursing Facilities*

Akin to hospitals are convalescent and nursing facilities, construction of which is increasing to meet growing demands for medical services for older persons.

These centers admit patients who require treatment less intensive but of longer duration than in a conventional hospital. Vertical transportation needs are less exacting in the convalescent home, which tends to be smaller than the intensive-care hospital and to have a smaller proportion of staff traffic but a relatively higher ratio of patient traffic.

Nursing homes shift the emphasis still further from active therapy to custodial care. In convalescent as well as nursing homes, vertical traffic can usually be handled by an elevator plant similar to that of an apartment building, except for the addition of at least one hospital-shaped service elevator for stretchers, wheelchairs, and other vehicles.

8.7. *Classroom Buildings*

Hospital vertical transportation systems are designed for the critical traffic periods when nurses and other staff members change shifts two or three times a day. In college and university classroom buildings, students changing classes place peak demands on elevator or escalator service as often as every hour.

High-rise buildings for higher education in the heart of larger cities made their appearance years ago. Today the goal of more intensive land utilization is bringing taller classroom, laboratory, and lecture-hall buildings even to outlying campuses.

In an undergraduate institution as much as 30 to 50 percent of the student body may be moving from floor to floor during the busiest 5 minutes of the time for changing classes. In the absence of measures to moderate the resulting traffic, it could overwhelm elevator service.

Building design can reduce the demand for vertical transportation by locating facilities like cafeterias and lounges, which draw heavy student traffic, on lower floors, and light-traffic facilities like administrative offices, higher up. Administrative procedures, such as those requiring students to walk up or down one flight during class change periods, also help. By reducing the number of elevator stops and round-trip time, such measures can maximize service for those who must use elevators.

For better handling of traffic surges, elevators should be wide and shallow, like those in department stores, with wide, center-opening doors. Multiple zoning for automatic group control may include provisions for continuous operation of cars during the busiest periods. Controls and fixtures should be designed to withstand possible abuse by excessively energetic students.

In academic buildings of eight or nine stories, escalators may well serve the heavy student traffic. Escalators, usually installed in pairs, may be reversed to provide all up service at the start of the first period, two-way service during the day, and all down service after the final period. A separate elevator is desirable for handicapped students and for service needs.

8.8. *Libraries and Museums*

Rather than being concentrated in periodic surges as in classroom buildings, vertical traffic in libraries and museums is continuous during the hours that the institution is open to users.

Traffic to a library reading room depends on its seating capacity and the average time each reader remains in the room. Circulation through a museum is also influenced by its capacity, and by the pace at which visitors view exhibits, but is difficult to estimate in advance.

Service for a library (Fig. 8.11) is usually provided by automatic elevators of the office-building type. Automatic dumbwaiters or

Fig. 8.11. Library of Auburn University (Alabama) is served by a pair of passenger elevators and a stack elevator.

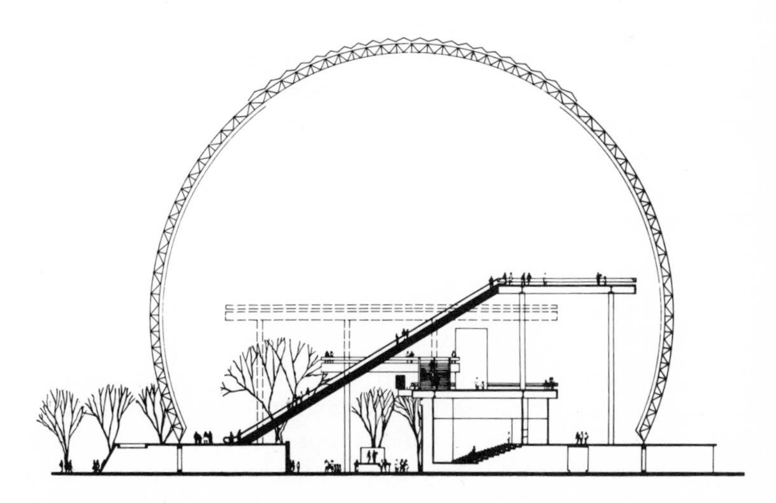

Fig. 8.12. Visitors at Expo 67 in Montreal rode North America's highest-rise escalator up the equivalent of six stories (68 ft. 7 in.) to view exhibits on platforms at different levels. The transparent geodesic sphere was designed by R. Buckminster Fuller, of Fuller and Sadeo, Inc, architect, and Geometrics, Inc, associated architect. Cambridge Seven Associates, Inc, who conceived the escalator system, were architects for the interior and exhibits.

service elevators may move books between stacks and reading room or circulation desk.

To accommodate the flow of visitors to upper-floor exhibit areas, escalators find increasing application in museums patronized by the public. In addition, one or two passenger elevators should be installed for handicapped persons and for special service, with a freight elevator for exhibits. Pavilions at major expositions increasingly use escalators, not only for transportation, but also to afford visitors striking views of exhibits (Fig. 8.12).

Elevators, escalators, and dumbwaiters are best planned as an entire system, all its elements working in unison to satisfy an institution's total vertical transportation needs. A successful installation can contribute materially to the quality and economy of the institution's service to its users.

Chapter 9

Industrial Buildings

After recent decades of preference for single-story plants, industry is once again awakening to the advantages of multilevel structures for manufacturing, distribution, and other operations. As with commercial, residential, and institutional buildings, rising land costs and, sometimes, the unavailability of additional ground space at any price are key factors in redirecting attention to the value of growing upward rather than spreading outward (Fig. 9.1).

Materials handling and transportation economics favor this new trend. Vertical movement in a multilevel plant eliminates longer, slower, and costlier horizontal handling in a sprawling one-story plant or over traffic-jammed roads. Many processes favor vertical handling, possibly with gravity flow from upper to lower floors.

Compact multilevel design also offers advantages in plant construction and operation. Excavation, foundation, and other building costs are often less per square foot of usable floor space, as are heating, cooling, and maintenance expenses. Specialized operations may be conducted on different floors, each with individually controlled conditions.

9.1. *Vertical Handling for Multilevel Plants*

Freight elevators are the most versatile means of moving materials in volume from level to level. In some plants an elevator permits intensive use of otherwise inaccessible basement or mezzanine areas. Designed for economical, fully automated handling of many kinds of materials, modern freight elevators are readily integrated with horizontal elements of plant-wide handling systems.

By facilitating materials handling throughout a plant, a well-designed elevator installation can make more floor space accessible and usable without requiring additional major construction. Fast, continuous vertical flow uses plant capacity more intensively for increased output at lower unit cost.

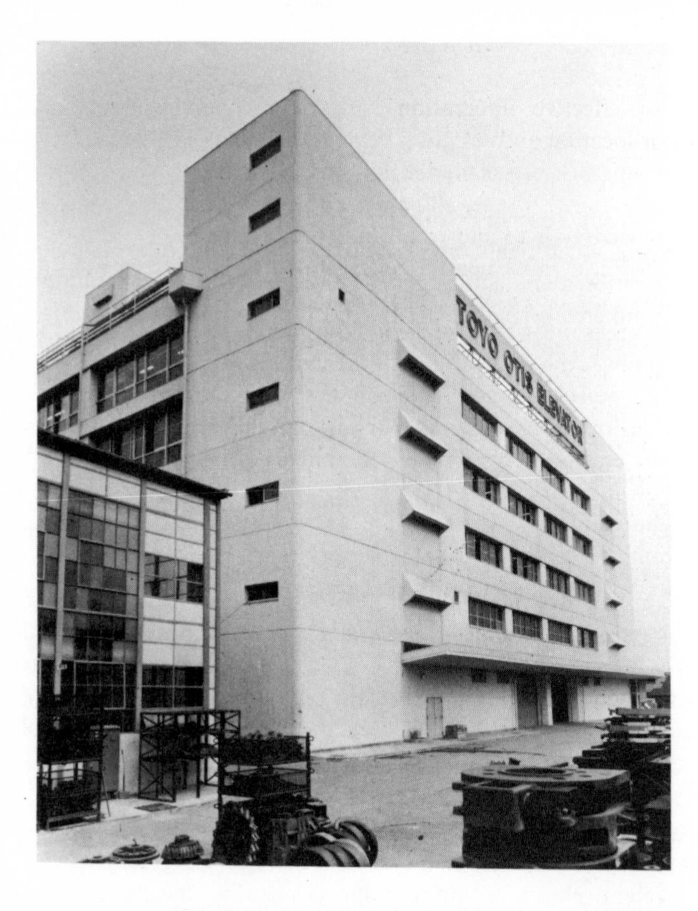

Fig. 9.1. Multilevel industrial buildings are advantageous under many conditions, and essential when ground space is limited. Freight elevators play a key role in materials handling in a multistory plant, like this one in Kamata, Japan.

To contribute most to production and materials handling economy, industrial-building elevators need the capacity and speed the particular plant requires. The installation must also be planned for long-run economy, flexibility to meet changes in products and methods, and dependability of operation and simplicity of maintenance.

9.2. *Elevator Location and Selection*

For effective integration with other plant facilities, freight elevator location and selection depend on analysis of the entire plant and its activities, especially the flow of materials. Relationship of areas for production, storage, shipping, and other operations and movement of materials to and from these areas influence elevator layout.

If vertical movement is concentrated in one part of the plant, elevators may be grouped there for more continuous service or, for several areas of heavy vertical flow, single elevators are spotted accordingly. An elevator may be located at the head of or alongside an aisle, or an elevator with doors at both ends may be placed between two aisles or even between two buildings, to serve both. If an elevator has two entrances, one may open on a loading platform and the other on interior areas of the plant.

Materials flow through the plant, especially in the vertical direction, determines the elevator work load. Its volume and pattern may be analyzed in terms of such factors as:

1. Materials to be carried, including weight, size, and shape of the largest loads.
2. Rate of vertical flow, in units, pounds, or tons per hour, shift or other period.
3. Distribution of vertical flow by floors and its fluctuation over a working day or other period.
4. Methods of horizontal materials handling—fork truck, conveyor, or other—with which elevators must be integrated.

Comprehensive analysis of the work load aids in selecting the type of elevator for an application; specifying driving machine, entrances, controls, and other elements of the installation; and determining capacity and speed. An elevator must have the dimensions and capacity to handle its bulkiest, heaviest load, possibly with a margin for future increases in the rate of production or the size or weight of products.

Fig. 9.2. Light-duty freight elevators have a self-supporting framework for installation in new or existing plants without reinforcing the building, adding overhead supports, or constructing a penthouse. Elevators with drum-type driving machines (illustrated) may be used for loads up to 2,500 lb. and rises to 35 ft.

9.3. *Types of Freight Elevators*

While plant layout influences elevator location, building structure and soil conditions also help determine the type selected. Freight elevators vary as widely as the work they do but are usually classed as light-duty, general-purpose, or industrial truck loading.

Light-duty freight elevators, for capacities up to 2,500 lb. and rises to 35 ft., may have drum-type traction or hydraulic driving machinery.

Light-duty drum-type elevators (Fig. 9.2) are usually the most economical to install. Machinery may be located in the basement or at a lower floor and needs no penthouse. Building structure requires little or no strengthening because the elevator's self-supporting guide rails transfer the weight of the car and its load to the pit.

For loads in excess of 2,500 lb., guide rails should be supported by the building structure. Typical dimensions for light-duty elevators appear in Fig. 9.3 and Table 9.1.

Table 9.1
Light-duty freight elevators, typical capacities and dimensions

	Platform		Hoistway		
Capacity (lb.)	(w) width	(d) depth	(W) width	(D) depth	Clear door width
1,500	5′ 4″	6′ 1″	6′ 11″	6′ 9″	4′ 5″
2,000	6′ 4″	7′ 0″	7′ 11″	7′ 8″	5′ 2″
2,500	6′ 4″	8′ 0″	7′ 11″	8′ 8″	5′ 2″

General-purpose freight elevators (Fig. 9.4) with capacities of 2,500 to 10,000 lb. meet a wide variety of handling requirements. Elevators of this type have either traction or hydraulic drive. Typical dimensions and capacities for traction-type general-purpose freight elevators appear in Fig. 9.5 and Tables 9.2 and 9.3, and in Fig. 9.6 and Table 9.4 for hydraulic elevators.

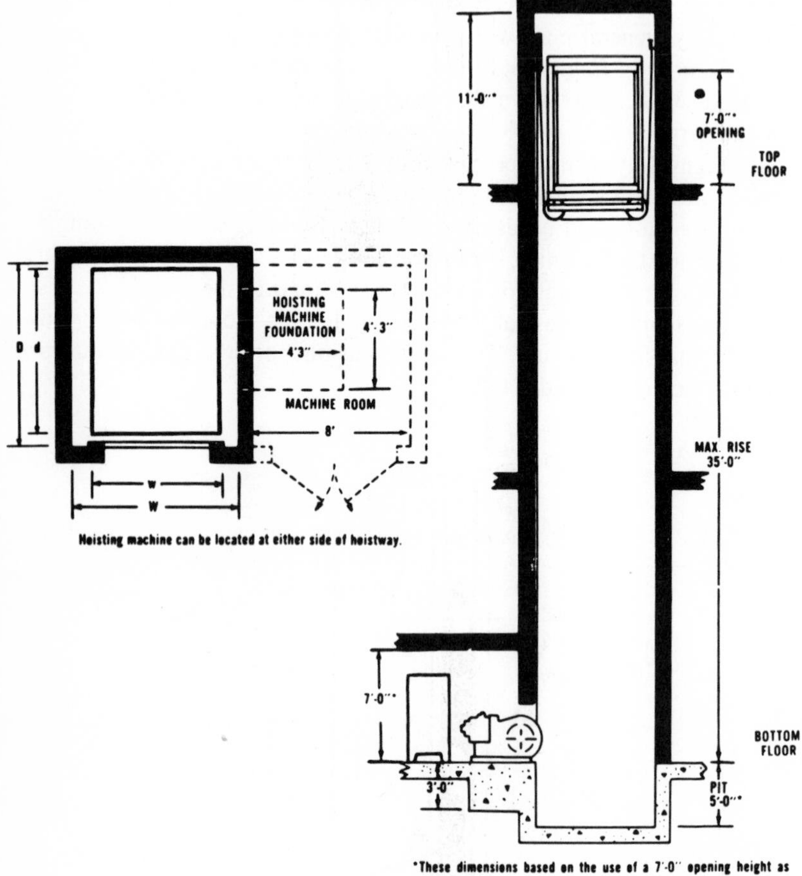

Fig. 9.3. Light-duty drum-type freight elevator, simplified elevation and plan. Hoistway, platform, and entrance dimensions appear in Table 9.1.

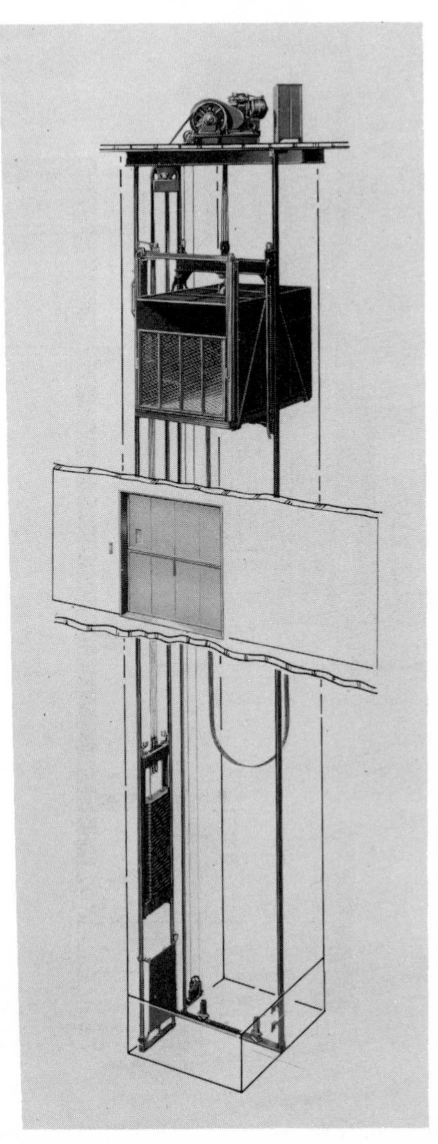

Fig. 9.4. General-purpose freight elevators have capacities of 2,500 to 10,000 lb. Traction-type (illustrated) may be installed in a building of any height. Modern hydraulic elevators are often used in plants of only a few stories.

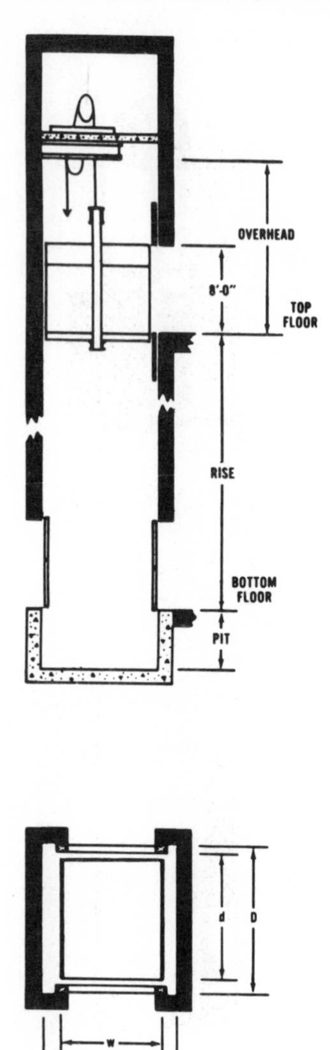

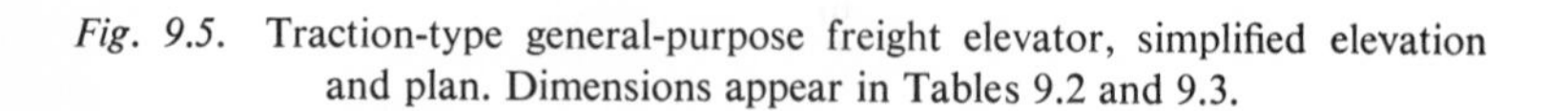

Fig. 9.5. Traction-type general-purpose freight elevator, simplified elevation and plan. Dimensions appear in Tables 9.2 and 9.3.

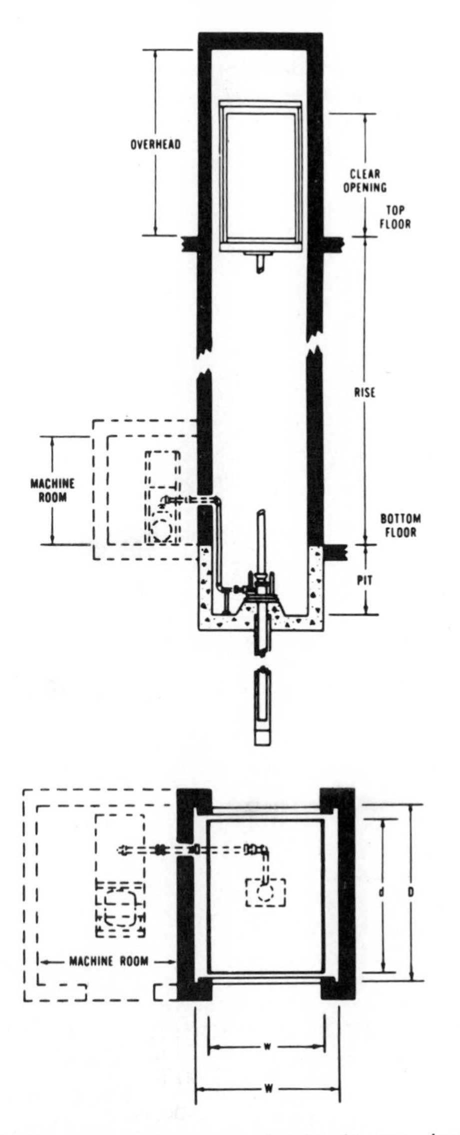

Fig. 9.6. Hydraulic-type general-purpose freight elevator, simplified elevation and plan. Typical capacities and dimensions appear in Table 9.4, but other combinations are available. Machine room dimensions vary with the size of equipment and building conditions.

Table 9.2
Traction-type general-purpose freight elevators, typical capacities and dimensions

Rated capacity (lb.)								Platform		Hoistway		Doors (stand. ht. 8′ 0″),
2,500	3,000	3,500	4,000	5,000	6,000	8,000	10,000	(w) width	(d) depth	(W) width	(D) depth	width
—	√	—	—	—	—	—	—	5′ 4″	7′ 0″	7′ 2″	7′ 11″	5′ 0″
√	—	√	√	—	—	—	—	6′ 4″	8′ 0″	8′ 4″	8′ 11″	6′ 0″
—	—	—	√	√	√	√	—	8′ 4″	10′ 0″	10′ 4″	10′ 11″	8′ 0″
—	—	—	—	—	—	—	√	8′ 4″	12′ 0″	10′ 6″	12′ 11″	8′ 0″

Table 9.3
Traction-type general-purpose freight elevators, minimum pit depths and overhead heights

	Capacity (lb.)							
	2,500		3,000		3,500		4,000	
Speed (fpm)	pit	o. ht.	pit	o. ht.	pit	o. ht.	pit	o. ht.
50–100	4′ 6″	14′ 5″	4′ 8″	14′ 9″	4′ 10″	14′ 11″	4′ 10″	14′ 6″
150–200	—	—	—	—	—	—	4′ 6″	14′ 6″
	5,000		6,000		8,000		10,000	
	pit	o. ht.	pit	o. ht.	pit	o. ht.	pit	o. ht.
45–100	4′ 10″	14′ 9″	4′ 10″	14′ 9″	4′ 10″	15′ 0″	4′ 10″	15′ 4″
150	—	—	4′ 8″	14′ 9″	—	—	—	—

Industrial truck freight elevators (Fig. 9.7, Table 9.5, and Fig. 9.8) with capacities of 10,000 to 20,000 lb. or more facilitate power truck handling on all floors of a plant. Special design and construction enable these elevators to take the severe stresses of impact, eccentric, and extra static loading that industrial trucks impose.

An industrial truck imparts impact loads when it rolls onto an elevator platform or brakes to a fast stop (Fig. 9.9). Eccentric loading

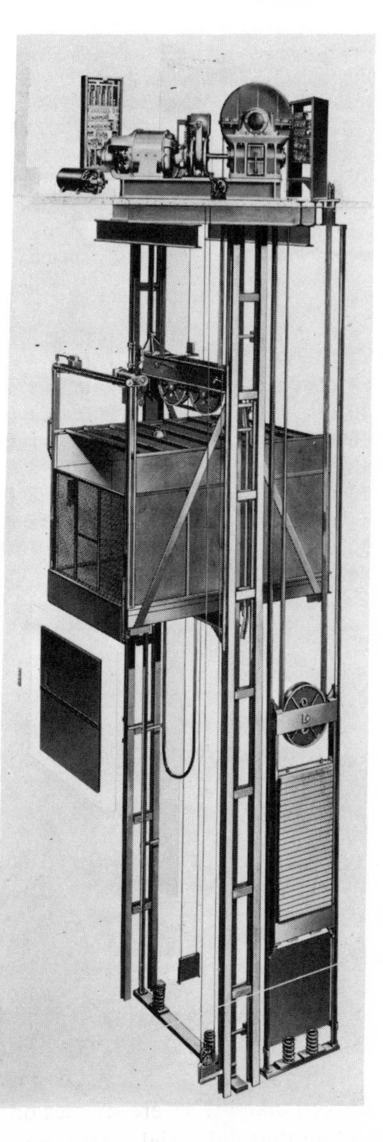

Fig. 9.7. Industrial truck elevators, built to take the exceptionally severe stresses of this form of loading, permit power truck handling on all floors of a plant. Capacities may range from 10,000 to 20,000 lb. Elevators may be traction (illustrated) or hydraulic type. Typical capacities and dimensions appear in Table 9.5.

Table 9.4
Hydraulic-type general-purpose freight elevators, typical capacities and dimensions

	Platform		Clear hoistway				
Capacity (lb.)	(w) width	(d) depth	(W) width	(D) depth	Doors, jamb opening	Pit	Over-head
2,500	5′ 4″	7′ 0″	6′ 8″	7′ 11″	4′ 3″ × 7′ 6″	4′ 3″	14′ 0″
3,000	6′ 4″	8′ 0″	7′ 8″	8′ 11″	6′ 0″ × 7′ 6″	4′ 3″	14′ 0″
3,500	6′ 4″	8′ 0″	7′ 8″	8′ 11″	6′ 0″ × 8′ 0″	4′ 6″	14′ 3″
4,000	6′ 4″	8′ 0″	7′ 8″	8′ 11″	6′ 0″ × 8′ 0″	4′ 6″	14′ 3″
5,000	8′ 4″	10′ 0″	9′ 10″	10′ 11″	8′ 0″ × 8′ 0″	4′ 6″	14′ 3″
6,000	8′ 4″	10′ 0″	9′ 10″	10′ 11″	8′ 0″ × 8′ 0″	4′ 6″	14′ 3″
8,000	8′ 4″	12′ 0″	9′ 10″	12′ 11″	8′ 0″ × 8′ 0″	4′ 6″	14′ 3″
10,000	8′ 4″	12′ 0″	11′ 4″	12′ 11″	8′ 0″ × 8′ 0″	4′ 10″	14′ 3″
12,000	8′ 4″	14′ 0″	11′ 6″	14′ 11″	8′ 0″ × 8′ 0″	5′ 0″	14′ 3″
16,000	8′ 4″	16′ 0″	11′ 8″	17′ 3″	8′ 0″ × 10′ 0″	5′ 6″	16′ 3″
18,000	10′ 4″	16′ 0″	13′ 8″	17′ 3″	10′ 0″ × 10′ 0″	5′ 6″	16′ 3″
20,000	12′ 0″	20′ 0″	15′ 4″	21′ 3″	11′ 8″ × 10′ 0″	5′ 6″	16′ 3″

Table 9.5
Industrial truck elevators; traction, typical capacities and dimensions

	Platform		Hoistway		
Capacity (lb.)	(w) width	(d) depth	(W) width	(D) depth	Doors, jamb opening
10,000	8′ 4″	12′ 0″	11′ 4″	12′ 11″	8′ 0″ × 8′ 0″
12,000	10′ 4″	14′ 0″	13′ 6″	14′ 11″	10′ 0″ × 8′ 0″
16,000	10′ 4″	14′ 0″	13′ 10″	15′ 3″	10′ 0″ × 10′ 0″
18,000	10′ 4″	16′ 0″	13′ 11″	17′ 3″	10′ 0″ × 10′ 0″
20,000	12′ 0″	20′ 0″	15′ 9″	21′ 3″	11′ 8″ × 10′ 0″

occurs as the truck's front wheels, on which the weight of truck and load is concentrated, roll onto the platform and tend to tilt the entire elevator forward.

Fig. 9.8. Hydraulic freight elevators for industrial truck handling may readily be installed in low-rise plants. The power unit may be located on the lowest level, next to or away from the hoistway.

Fig. 9.9. As the fork truck operator drives onto an elevator, the moving, loaded truck exerts impact forces that tend to thrust, tilt, and twist the entire elevator structure. Weight of truck and load on one part of the platform tends to unbalance the elevator. Industrial truck elevators are built to resist these impact forces and unbalanced loads.

Extra static loading occurs as the truck deposits its final load on the elevator, which must then support not only its rated live load but also the weight of the truck (Fig. 9.10). To resist these stresses, industrial truck elevators have heavy-duty hydraulic or traction driving machines. Cars and hoistway structures are of high-strength design and construction, and the building structure is reinforced for added rigidity.

Freight elevators with capacities exceeding 100,000 lb. have been installed to carry loaded highway trucks and trailers directly to upper

Fig. 9.10. A fork truck may place load after load on the elevator until it is full. As the truck deposits its final load, the elevator may have to support its full pay load plus the weight of the truck on its front wheels. Industrial truck elevators can take this extra static load, which may equal 50% of rated capacity.

levels of plants. These capacious, powerful units permit unloading an over-the-road vehicle near where its contents will be used, eliminating transfer to in-plant vehicles.

For explosive atmospheres and other hazardous locations, elevators are installed with special hoistway equipment that will function safely and dependably under service conditions. The machine room can be located outside the danger area, or sealed or otherwise isolated from it. Explosionproof machines and controllers may be specified where necessary.

For simplified elevations and plan of freight elevators for industrial truck handling see Fig. 9.11.

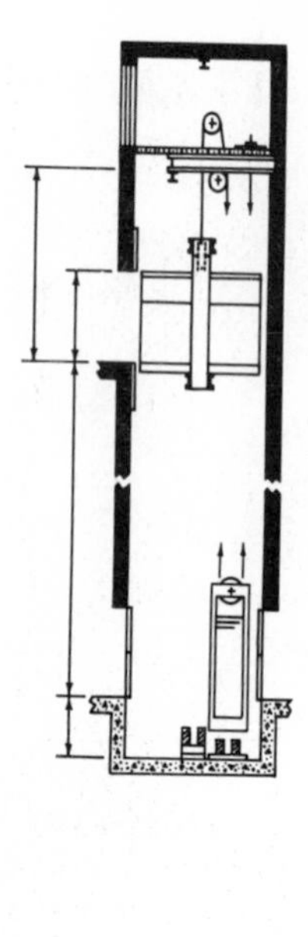

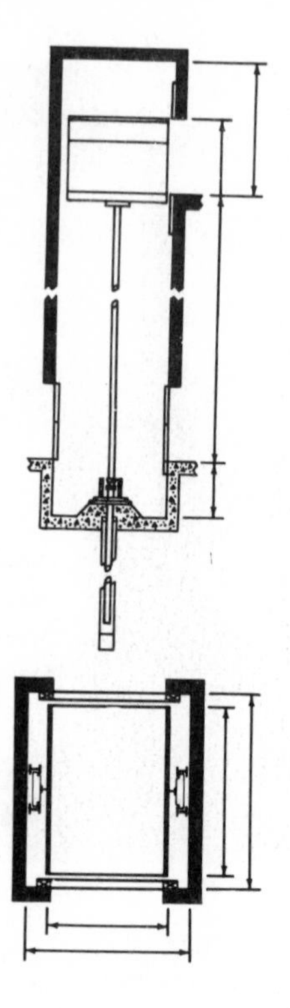

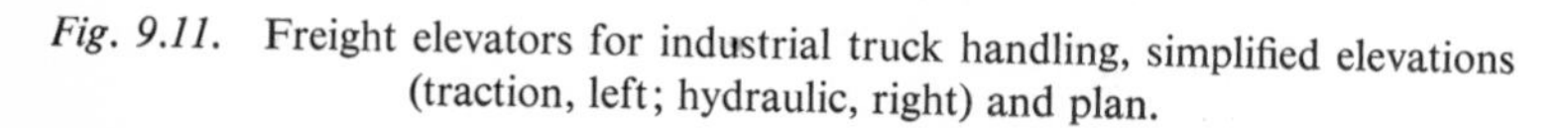

Fig. 9.11. Freight elevators for industrial truck handling, simplified elevations (traction, left; hydraulic, right) and plan.

9.4. *Cars and Entrances*

Hoisting ropes or a hydraulic cylinder apply lifting action to the elevator car through the car frame and platform, which are made of structural steel to take the forces of freight elevator service. Mounted on the platform is the car enclosure of steel panels interlocked so that all joints are flush. Control panels and car lights are recessed to protect them from damage by moving freight or vehicles.

Elevator entrance design and operation influence the speed with which the car can be loaded and unloaded and the efficiency and

Fig. 9.12. Hoistway entrances influence plant design and speed and safety of floor-to-floor handling. Biparting hoistway doors (shown) open to leave the entire car width clear for loading. Power operation opens doors automatically as the elevator levels to a stop.

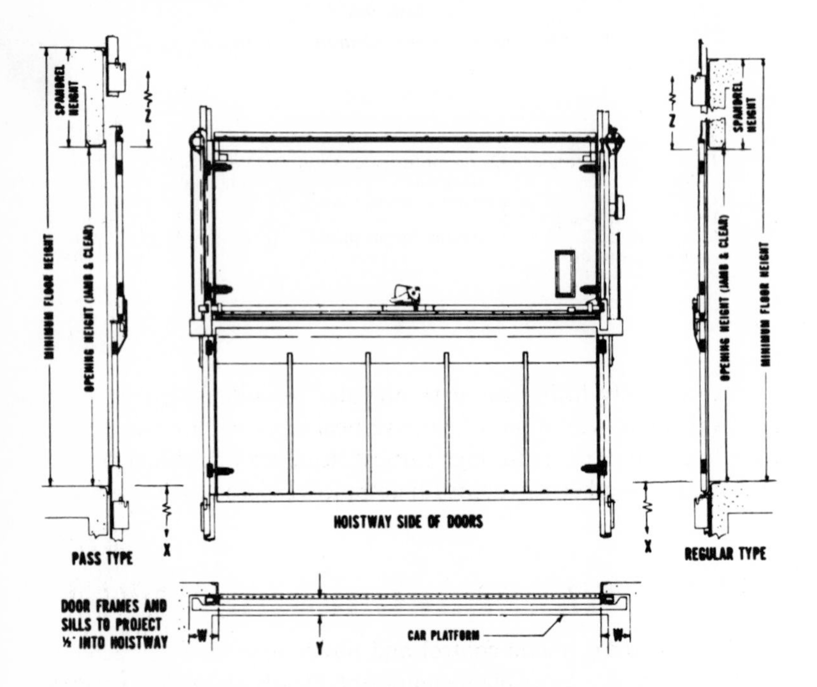

Fig. 9.13. Biparting hoistway doors, pass and regular type. For pass-type doors, spandrel height must be at least 15½ in.; for regular type, spandrel height must be at least one-half opening height plus 7½ in. Clearance dimensions appear in Table 9.6.

safety of the entire installation. Hoistway doors of modern biparting design (Figs. 9.12 and 9.13 and Table 9.6), with two counterbalanced vertically sliding sections, open to leave the car's full width free for loading.

When doors are open, the lower section forms a bridge to the car platform, eliminating need for a landing sill. All-steel doors are light, strong, and fire-resistant. Hoistway doors and car gates should be power-operated.

Table 9.6
Biparting hoistway doors, clearance dimensions

		Regular type	Pass type
W	Jambs, manual door	9″	9″
	Jambs, power-operated door	12″ on motor side, 9″ on other side	12″ on motor side, 9″ on other side
X	Pit depth	½ opening height plus 3″	½ opening height plus 3″
Y	Hoistway sill to car	5″	6⅝″
Z	Clearance above opening on top terminal floor	½ opening height plus 4″	½ opening height plus 4″

A vertically lifting car gate operates in conjunction with the hoistway doors. Variations of basic vertical biparting hoistway doors include pass-type doors for high vertical clearance where floor heights are low, and weatherstrip doors for entrances exposed to the elements.

9.5. *Types of Control*

Trends to automatic control and power operation characterize elevators as well as other plant equipment. Nearly all freight elevators installed in recent years operate automatically, and older, manually operated elevators are often modernized for automatic control.

Operating economy is a prime objective of elevator automation. Another is full-time availability, whether or not a skilled attendant is on hand. Finally, automatic control systems can improve elevator efficiency and substantially increase effective handling capacity.

Reducing round-trip time lets an elevator do more work. In a two- or three-story plant, opportunities for reducing round-trip time by increasing elevator speed are limited. But seconds can be saved at every stop when elevator starting, stopping, acceleration, deceleration, and leveling are automatically controlled and coordinated with opening and closing of power-operated entrances. Automatic controls perform these operations with speed, precision, and safety.

Several systems of automatic operation are available, each applicable to the principal types of freight elevators.

Double-button operation may be used for light- to moderate-duty elevators. A pair of buttons, "up" and "down," are mounted in the car and at the landings. The elevator moves up or down as long as the appropriate button is pressed. The car control panel may also have an "inching" button to move the car a short distance with its doors open and bring it level with a landing.

Single automatic operation (Section 5.3) should be considered for an elevator in locations where a single load is usually large enough to fill the car. An elevator with this form of automatic control answers only one call on each trip, permitting uninterrupted use of the elevator until it completes a trip. With this system's "call-and-send" variation, the elevator can be controlled entirely from landing buttons or special automatic controls integrated with powered conveyors (Section 9.6), so that no one need enter the car.

Automatic leveling, which is desirable for any busy freight elevator, is virtually essential where elevators are loaded by manual or power trucks. Truck wheels should be of sufficient diameter to bridge easily the standard $1\frac{1}{4}$-in. running clearance between elevator platform and door sill.

Power operation of elevator entrances speeds opening and closing, saving additional time by beginning to open doors and gates automatically as the car levels to a landing. Elevators with entrances at both ends may be equipped for selective (doors opening only at the end where a button is pressed) or nonselective operation (both doors opening if a button at either end is pressed).

9.6. *Horizontal-Vertical Handling*

Efficient automatic elevator and door operation, by reducing round-trip time, increases an elevator's effective handling capacity. Faster loading and unloading, another way to get more work from a freight elevator, depends on close integration of the elevator installation with the rest of the plant-wide handling system.

Elevator location, size, shape, and entrance arrangements should be related to the building plan. Some plant layouts, for instance, permit accumulating loads at landings to keep the elevator working rather than waiting.

Loading and unloading speed also depends on the loading

method: manual, hand, or power truck or conveyor. Where industrial trucks are used, they need room at landings to maneuver. Pushbutton landing fixtures suspended from the ceiling along approaches to the elevator permit a driver en route to a landing to call the automatic elevator without dismounting (Fig. 9.14).

Fig. 9.14. To speed materials handling by fork truck and freight elevator, landing button fixtures may be hung from the ceiling along approaches to the elevator. Pendant controls let the driver call the elevator without dismounting.

Loading or unloading may be automated in various ways, depending on materials and methods of handling. In a newspaper publishing plant, for example, delivering rolls of newsprint from level to level is facilitated by automatic freight elevators with power-operated tilting platforms.

To unload at a landing, the elevator platform tilts upward at the back just enough for the paper to roll out the front of the elevator. The platform then retires to its normal position, the doors close and the elevator travels to another level. Its doors open automatically and the elevator is ready for its next load, the complete cycle taking about 2 minutes. When the elevator carries general freight, the tilting mechanism does not operate.

9.7. *Automated Handling Systems*

Progress in integrating automated freight elevator and powered conveyor systems points the way to compact, multilevel plants in which materials move as readily as on a single floor. Such installations may incorporate powered conveyors in the elevator as well as at the landings with integrated control systems to coordinate automatically the operation of elevator and conveyors (Fig. 9.15).

In these plants, pressing a button suffices to dispatch loads from locations on one level to other levels and locations. Under direction of the integrated control system, floor conveyors transfer a load to the elevator which rises or descends to the desired level, where the conveyor on the elevator automatically discharges the load onto another floor conveyor for horizontal movement. Automatic controls may also be set to repeat continuous cycles of delivering and returning loads to and from various floors and stations.

Merchandise transfer between floors of multistory warehouses may be automated by horizontal-vertical handling systems that integrate modified freight elevators with carts moved by underfloor towlines (Fig. 9.16). Personnel simply load merchandise onto carts, set probes for desired destinations, and push carts onto towline spurs. Automatic equipment then performs all operations: transfer of carts to and onto elevators, transportation between floors, and transfer from elevators to unloading areas.

Loaded carts, with destination probes set, are pulled by the subfloor towline to the elevator area and switched onto the spur leading to the elevator serving the probed destination. Each spur has two tracks and each track takes up to three carts. If both tracks of a spur are filled, a cart continues to circulate on the main loop until space is available.

Fig. 9.15. Automated conveyor-elevator system moves palletized loads between receiving platforms and points on upper floors. The right-angle transfer unit, with power-operated chains and rollers, delivers loads to the elevator, which has a power roller conveyor to move them on and off the car. The gate closes and the elevator starts automatically when electric eyes determine that a load is properly positioned on the platform.

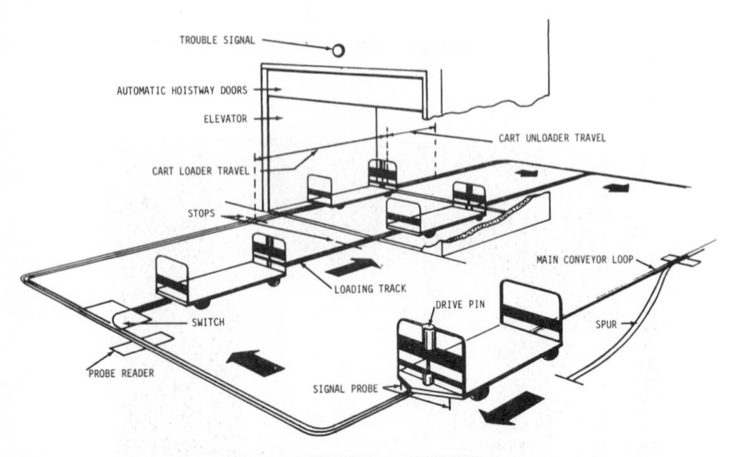

Fig. 9.16. Interfloor merchandise handling is automated by system of freight elevators integrated with carts and underfloor towline conveyors. Carts are automatically directed to desired destinations by signal probes that operate track switches and elevator controls.

Arrival of a cart on a spur automatically calls the elevator to that floor. By their positions, the destination probes "instruct" the automatic elevator to go to the floor to which the car is to be sent. When the elevator arrives, transfer mechanisms in the elevator floor pull a cart in and, at the desired floor, push it out onto the floor-wide towline system.

Systems developed initially to park automobiles completely automatically in multistory buildings (Section 10.9) also hold promise for automated, integrated horizontal-vertical materials handling. A special freight elevator (for vertical movement) is installed in a traveling tower (for horizontal movement) that runs on tracks between rows of stalls on many levels in which automobiles or palletized loads can be stored (Fig. 10.16). In a variation (Fig. 9.17) of this system the elevator travels only vertically, but often to greater heights, automatically transferring loads to storage spaces one or two rows deep on either side of the hoistway.

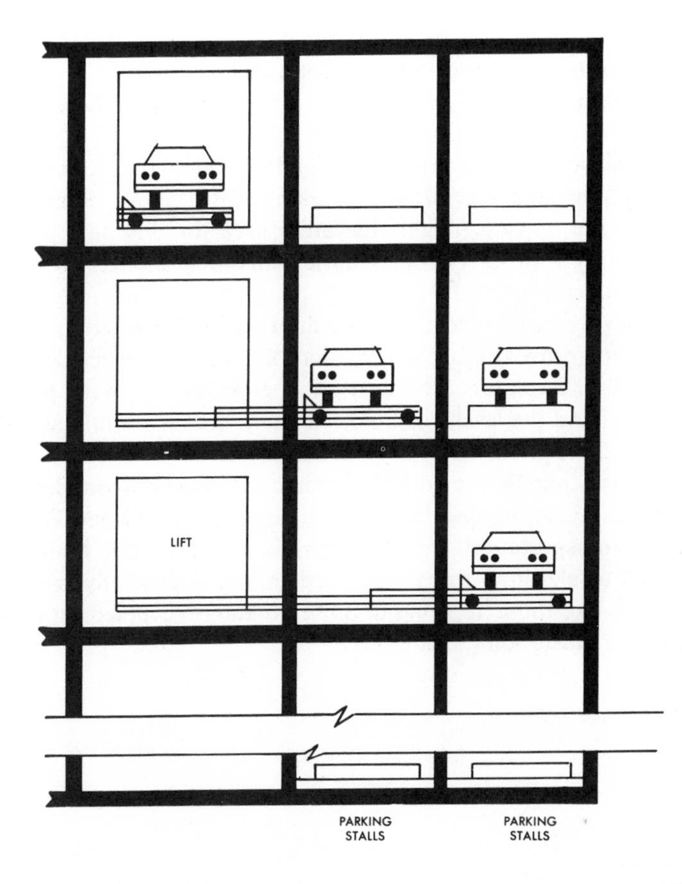

Fig. 9.17. Automated parking facility for automobile assembly plant at St. Thomas, Ontario. Automobiles awaiting final work are automatically stored and retrieved two deep in stalls on both sides of special elevator. Telescoping fingers on lift place cars in and out of stalls next to and farther from hoistway.

On the elevator platform, steel fingers operating much like the familiar industrial truck's lift fork pick up a load and gently lift it on or off the elevator at a storage space or unloading station. Inserting a data processing-type punchcard in a special console automatically controls the entire operation, delivering or retrieving the particular load desired.

9.8. *Elevators and Escalators for Personnel*

In addition to vertical handling of material, a multilevel plant or warehouse may also require floor-to-floor transportation for its personnel.

Employees in many industrial buildings ride freight elevators at the beginning and end of each working shift, before or after the elevators are moving freight. Elevators used for this dual purpose must comply with safety code requirements for passenger as well as for freight service.

Carrying people from floor to floor while freight elevators are busy at their primary job may necessitate separate passenger elevators. Increasingly, escalators are transporting employees between production floors and the levels on which cafeteria, locker room, and other facilities are located.

Able to handle the surges of traffic that occur when shifts are changing, escalators find favor especially in larger plants. Whether vertical transportation for passengers in an industrial building is assigned to elevators or escalators, planning is generally based on principals like those applying to office buildings occupied entirely by a single organization (Section 6.8).

Chapter 10

Buildings of Special Design

The function or use of a building dominates the planning of its vertical transportation system. Because each principal type of building has its distinctive requirements for service, the four preceding chapters have outlined characteristics of elevator and escalator systems for commercial, residential, institutional, and industrial structures.

But requirements depend not only on a building's function but also on its design, particularly its height and shape. This chapter therefore introduces concepts peculiar to elevatoring low-rise, sheer-rise, and very tall buildings. In-building parking and multipurpose buildings also call for special provisions in vertical transportation, reviewed in the final pages of the chapter.

10.1. *Low-Rise Buildings*

Low-rise buildings present special problems in planning elevator installations to satisfy architectural and other design requirements. Whether the building is occupied by owner or tenants or both, prompt, responsive elevator service may be an economic necessity. Total elevator costs, immediate and ultimate, must be held to a minimum which is consistent with desired service standards.

Modern hydraulic elevators, operating automatically, with electrically powered pumps and oil as the working fluid, have been widely installed in low-rise buildings. Higher-speed "low-overhead" electric traction elevators are now also being specified for buildings of moderate height with low roof lines.

Building design and construction impose limits on the elevator installation. Modern low-rise buildings often have flat roof lines that rule out high penthouses required for conventional traction elevators. Subsurface conditions, on the other hand, may make drilling for a hydraulic cylinder unduly expensive.

10.2. *Hydraulic Elevators*

Installed as a rule with its pumping unit in the basement, a hydraulic elevator eliminates the rooftop penthouse. Since the elevator is raised and lowered by a plunger moving in a cylinder sunk in the ground, vertical stresses on the hoistway structure are negligible and heavy framing is usually not required.

Elevator components have long been standard, but complete equipment for each job have had to be engineered and manufactured to individual specifications. Now, however, the principle of pre-engineering is being applied to the entire hydraulic elevator. Ordering and specifications have also been standardized, contributing to substantial savings in total costs.

Redesigned structural elements eliminate excess weight and bulk, yet retain their essential strength. Resulting lighter loads on the hydraulic drive machinery permit use of motors, pumps, and other components of lower rating without impairing performance.

Power units incorporate pump, motor, and controller in compact, self-contained assemblies which may be installed wherever space is available. Quality elevator service may now be provided in apartments and other types of low-rise buildings where it had been considered uneconomic (Fig. 10.1).

Pre-engineered hydraulic elevators are usually of 1,500 and 2,000 lb. capacity (Fig. 10.2). Control is collective automatic with two-way self-leveling and power-operated car and hoistway doors.

Acceptance of pre-engineered hydraulic elevators is bringing application of the principle to low-rise, light-duty traction elevators. With an increasing number of elderly persons and rising expectations of service by people of all ages, pre-engineering for hydraulic or traction elevators appears a timely innovation.

10.3. *Wheelchair Lifts*

Since even short flights of stairs bar disabled persons in wheelchairs, an economical lift has been developed to make low-rise institutional and residential buildings more accessible to the handicapped. Former "architectural barriers" of this nature are overcome by the

Fig. 10.1. Low-rise apartment building served by pre-engineered hydraulic elevator of standardized design. Cornwallis Manor, Greensboro, N.C.; Adrian P. Stout, architect; Randall Sheppard, general contractor.

wheelchair lift, designed for ready installation on the outside of existing buildings.

To save costs, the car, platform, and car frame are of unitized construction, light but sturdy. The car is raised and lowered by a telescoping hydraulic lift somewhat like those of industrial fork trucks.

Installed above ground, this system eliminates the hole that must be drilled for the conventional hydraulic elevator. A simple shaftway for the lift, designed to blend with the architecture, may be constructed outside the building (Fig. 10.3). Operating automatically and leveling itself at each stop, the lift carries a wheelchair paraplegic and his attendant.

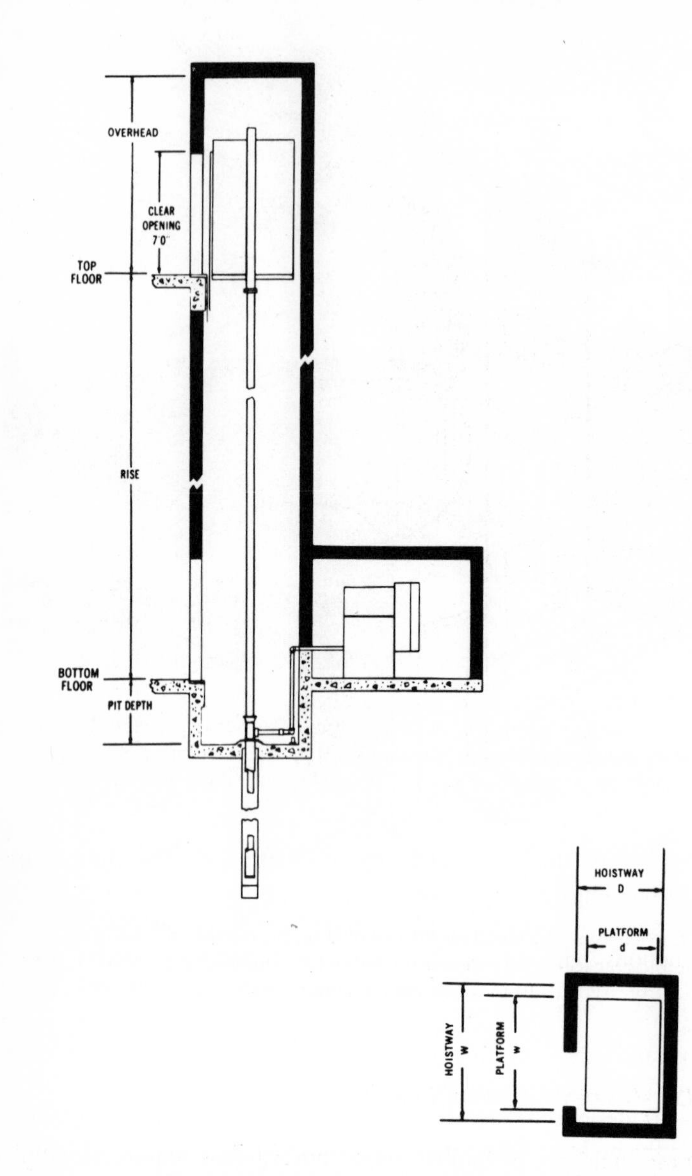

Fig. 10.2. Simplified elevation and plan of pre-engineered hydraulic elevator. Dimensions depend on elevator capacity.

Fig. 10.3. Economical wheelchair lift installed on the outside of a building. Car has front and rear gates, for access from without the building at ground level and from within on the upper floor.

10.4. *Low-Overhead Traction Elevators*

Many buildings of limited height nevertheless require elevator speeds as great as 350 fpm for adequate service. While traction elevators can be operated at these higher speeds, conventional arrangements, with driving machinery over the hoistway, require high penthouses and are unsuited to buildings with low roof lines. Types have therefore been engineered with special roping arrangements to permit locating the machine in the basement (Figs. 10.4 and 10.5).

These "low-overhead" traction elevators retain the inherent operating and safety characteristics of this type of equipment but eliminate the usual penthouse machine room. Compact driving machines may be located alongside the hoistway at the lower landing.

One roping arrangement, with an underslung car, requires a total height of as little as 12 ft. 11 in. above the top floor (Fig. 10.6, Table 10.1). A second arrangement, with overhead suspension, reduces pit depth but requires slightly more height above (Fig. 10.7, Table 10.2). With either arrangement, the hoistway roof is so low that it usually does not project above the building roof or parapet.

Future construction of additional upper stories is simplified by locating the driving machine at the lower landing. It is unnecessary either to raise the machine, as would be required for a conventional, overhead traction elevator, or to drill a deeper hole, as for a hydraulic elevator.

Table 10.1

Duties and dimensions, traction elevators for minimum overhead height

Duty		Hoistway					Platform				
Load (lb.)	Speed (fpm)	A	B	Door space, C	Doors, clear opening	Clear to sill, D	E	F	Mach. room, G	Min. pit depth	Min. overhead height
1,200	100	6′ 9″	5′ 8½″	6½″ 2 speed	3′ 0″	5′ 2″	5′ 0″	4′ 0″	8′ 6″	5′ 9″	12′ 11″
2,000	100, 150	8′ 1½″	6′ 1″	5″ single slide	3′ 0″	5′ 8″	6′ 4″	4′ 5″	10′ 0″	5′ 9″	13′ 0″
	200	8′ 3″	6′ 3″	5″ single slide	3′ 0″	5′ 10″	6′ 4″	4′ 5″	11′ 0″	6′ 4″	13′ 11″
	250, 300	8′ 3½″	6′ 3″	5″ single slide	3′ 0″	5′ 10″	6′ 4″	4′ 5″	11′ 0″	6′ 8″	13′ 11″
2,500	100, 150	8′ 5½″	6′ 8″	5″ single slide	3′ 0″	6′ 3″	7′ 0″	5′ 0″	10′ 0″	5′ 9″	13′ 0″
	200	8′ 11″	6′ 10″	5″ center opening	3′ 6″	6′ 5″	7′ 0″	5′ 0″	11′ 0″	6′ 5″	13′ 11″
	250, 300	8′ 11½″	6′ 10″	5″ center opening	3′ 6″	6′ 5″	7′ 0″	5′ 0″	11′ 0″	6′ 8″	13′ 11″
3,000	200	8′ 11½″	7′ 4″	5″ center opening	3′ 6″	6′ 11″	7′ 0″	5′ 6″	11′ 6″	6′ 8″	14′ 1″
	250, 300	8′ 11½″	7′ 4″	5″ center opening	3′ 6″	6′ 11″	7′ 0″	5′ 6″	11′ 6″	7′ 0″	14′ 1″
3,500	200	8′ 11½″	8′ 0″	5″ center opening	3′ 6″	7′ 7″	7′ 0″	6′ 2″	11′ 6″	6′ 9″	14′ 3″
	250, 300	8′ 11½″	8′ 0″	5″ center opening	3′ 6″	7′ 7″	7′ 0″	6′ 2″	11′ 6″	7′ 0″	14′ 3″
4,000 (Hospital)	75	8′ 0½″	9′ 6½″	6½″ 2 speed	4′ 0″	9′ 0″	5′ 8″	8′ 8″	6′ 8″	6′ 2″	14′ 10″

Fig. 10.4. "Low-overhead" passenger elevator for moderate-height building with low roof line. Electric traction driving machine, at lower landing, retains inherent operating and safety characteristics of this type of equipment while simplifying building structural requirements.

Table 10.2
Duties and dimensions, traction elevators for minimum pit depth

Duty		Hoistway					Platform				
Load (lb.)	Speed (fpm)	A	B	Door space, C	Doors, clear opening	Clear to sill, D	E	F	Mach. room, G	Min. pit depth	Min. overhead height
1,200	100	6′ 5″	5′ 9½″	6½″ 2 speed	3′ 0″	5′ 3″	5′ 0″	4′ 0″	8′ 6″	4′ 10″	15′ 4″
2,000	100 150	7′ 9½″	6′ 1″	5″ single slide	3′ 0″	5′ 8″	6′ 4″	4′ 5″	10′ 0″	4′ 9″	15′ 10″ 16′ 1″
	200 250 300	7′ 9½″	6′ 1″	5″ center opening	3′ 0″	5′ 8″	6′ 4″	4′ 5″	11′ 0″	5′ 0″	16′ 0″
2,500	100 150	8′ 5½″	6′ 8″	5″ single slide	3′ 6″	6′ 3″	7′ 0″	5′ 0″	10′ 0″	4′ 9″	15′ 10″ 16′ 1″
	200	8′ 5½″	6′ 8″	5″ center opening	3′ 6″	6′ 3″	7′ 0″	5′ 0″	11′ 0″	4′ 9″	16′ 0″
	250 300	8′ 5½″	6′ 8″	5″ center opening	3′ 6″	6′ 3″	7′ 0″	5′ 0″	11′ 0″	5′ 0″	16′ 2″
3,000	200	8′ 9″	7′ 4″	5″ center opening	3′ 6″	6′ 11″	7′ 0″	5′ 6″	11′ 6″	4′ 9″	16′ 6″
	250 300	8′ 9″	7′ 4″	5″ center opening	3′ 6″	6′ 11″	7′ 0″	5′ 6″	11′ 6″	5′ 0″	16′ 6″
3,500	200	8′ 7″	8′ 0″	5″ center opening	3′ 6″	7′ 7″	7′ 0″	6′ 2″	11′ 6″	4′ 10″	16′ 8″
	250 300	8′ 7″	8′ 0″	5″ center opening	3′ 6″	7′ 7″	7′ 0″	6′ 2″	11′ 6″	5′ 1″	16′ 8″
4,000 (Hospital)	200	8′ 0″	9′ 6½″	6½″ 2 speed	4′ 0″	9′ 0″	5′ 8″	8′ 8″	11′ 0″	5′ 1″	16′ 8″

One or two low-overhead elevators are controlled by automatic systems of the usual collective type. Groups of three or more of these elevators are coordinated by multiple zoning systems similar to those in taller buildings with conventional, overhead traction machines.

10.5. *Sheer-Rise Towers*

As buildings become taller, their design, in height, bulk, or shape, becomes increasingly subject to local zoning laws. While earlier regulations attempted to assure light and air by specifying setbacks, more recent zoning laws require leaving a portion of the site clear but permit the building itself to rise straight to a greater height. Advantages include a high proportion of outside space and compact floor plans with short, direct, horizontal lines of communication.

In requirements for elevator service, sheer-rise towers differ significantly from setback buildings. Successive setbacks reduce gross areas of upper floors, usually by amounts greater than areas gained as elevator hoistways terminate. Net floor areas therefore become progressively smaller in upper parts of setback buildings, with population per floor and the demand for elevator service following the same trend.

Each elevator in a group serving a building's upper floors delivers less service, in terms of quantity and quality, than an elevator of the same rating operating to lower floors. Despite its higher speed, each high-rise elevator's effective carrying capacity is reduced by long runs to and from upper floors.

Fig. 10.5. Electric elevators of low-overhead type permit flat roof line of Center for Materials Science and Engineering at Massachusetts Institute of Technology, Cambridge, Massachusetts. Skidmore, Owings and Merrill, architect; George A. Fuller Company, general contractor.

This decrease in the handling capacity of upper-floor elevators can readily be matched, in setback buildings, to the progressive decrease in population and traffic demand. But the very opposite may prevail in sheer-rise towers.

Gross area per floor is uniform from top to bottom of the sheer-rise building. But the terminating of lower-floor elevator groups

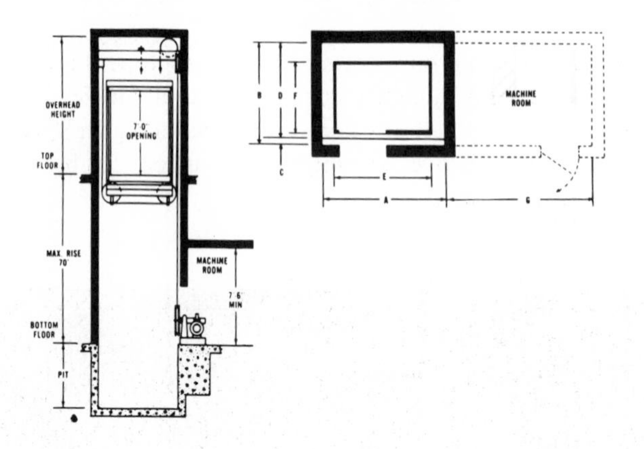

Fig. 10.6. Machine-below traction elevator arrangement for minimum overhead height. Plan dimensions refer to Table 10.1.

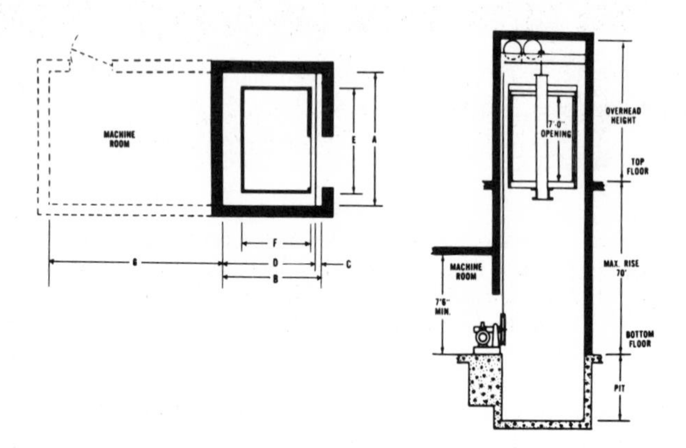

Fig. 10.7. Machine-below traction elevator arrangement for minimum pit depth. Plan dimensions refer to Table 10.2.

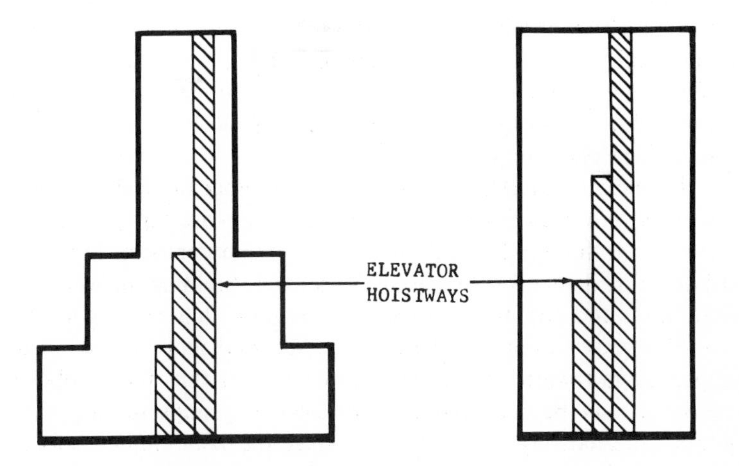

Fig. 10.8. In setback buildings (left), net area per floor decreases with increasing height. But upper floors of sheer-rise towers (right) have greater net areas as elevator hoistways terminate. With heavier rather than lighter traffic to upper floors, elevators serving them must be engineered for greater handling capacity despite longer round-trip time.

results in greater net areas for upper floors, which may thus have the larger populations and heavier traffic (Fig. 10.8).

10.6. *Satisfying Service Needs*

Other factors also influence elevator service requirements. To produce the same net floor area as the setback building, the slender tower must be taller. Tower space enjoys inherent advantages that attract tenants paying higher rentals and expecting superior service.

Choice upper floors devoted largely to managerial or professional personnel may not account for a great volume of elevator traffic but do require service of high speed and prompt response. The value of single-floor layouts, on the other hand, may lead firms to devote entire floors in the tower to clerical areas that generate heavy peak traffic.

Because of the tower's height and separation from neighboring structures, its roof may prove a desirable location for a restaurant or other facility. A rooftop restaurant provides a welcome service for a building's occupants and visitors—and additional earnings for its owner—but imposes demands for elevator service deserving special attention in planning the tower.

In many setback buildings the same number of elevators serves the lower, intermediate, and upper floors. High-rise cars provide less service than those to lower floors, a pattern that conforms to the distribution of demand throughout the setback building, in which net floor areas and population per floor decrease with ascending heights.

Sheer-rise towers, on the other hand, need more service to the upper floors than to the lower. To accomplish this, the number of elevators may be progressively increased in groups serving upper floors. Alternatively, all groups may have the same number of cars, but the high-rise elevators will serve the smallest number of floors; the low-rise, the largest number of floors; and the intermediate, an intermediate number.

Elevator service to upper floors can be improved by increasing express-run speeds to minimize transit time. Quickening the pace of service while reducing its operating cost, elevator automation also contributes to the economic feasibility of the sheer-rise tower. Shallow cars with wide, center-opening doors let people in and out faster, reduce the time from one stop to the next, and increase the number of stops each elevator can serve in a given period.

10.7. *Very Tall Buildings*

When towers attain heights as great as those of Chicago's 100-story John Hancock Center (Fig. 10.9), the twin 110-story tower buildings (Fig. 10.10) of the World Trade Center in New York, and possibly buildings of even loftier stature, vertical transportation becomes more critical than ever. Passengers to upper portions of the tower need fast, frequent, high-capacity elevator service, yet hoistways must not consume so much space on lower floors as to make the building uneconomic.

Architects and engineers have therefore sought ways to use

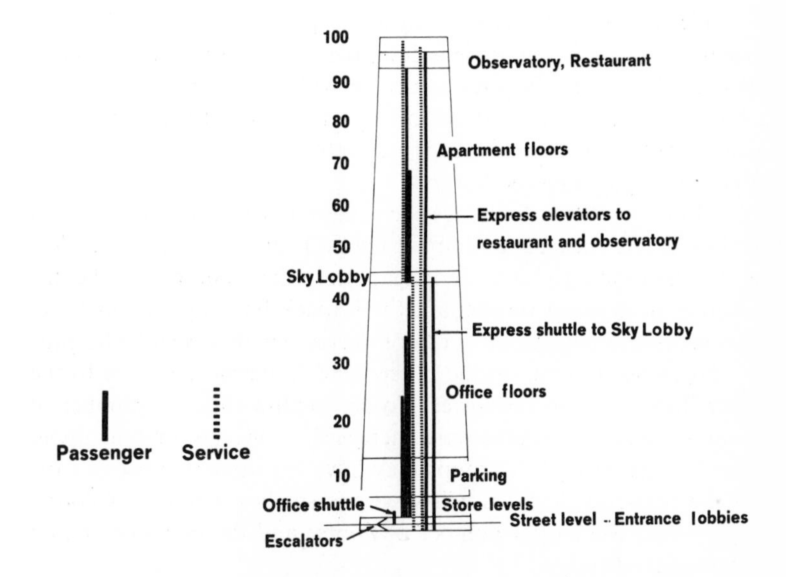

Fig. 10.9. Elevatoring buildings of very great height becomes practical with the sky lobby concept, here used in the 100-story John Hancock Center in Chicago. Apartment residents take shuttle elevators to sky plaza level, express and local elevators from there up to apartment floors. Elevator system with stacked hoistways saves lower-floor space, yet provides high-quality service to all parts of the building. Skidmore, Owings and Merrill, architect-engineer; Tishman Construction Company, general contractor.

hoistways more intensively, thus providing more transportation service for the building space consumed. Research and development effort have embraced such means as higher elevator speeds, sky lobby systems, double-deck elevators, and several cars in a single hoistway.

10.8. *Increased Travel Speed*

As doubling the speed of jet liners from a subsonic 600 mph to a supersonic 1,200 roughly doubles the passenger-miles a fleet can generate in an hour or a day, raising elevator speeds increases the service from the same number of cars and hoistways. Actual gains in

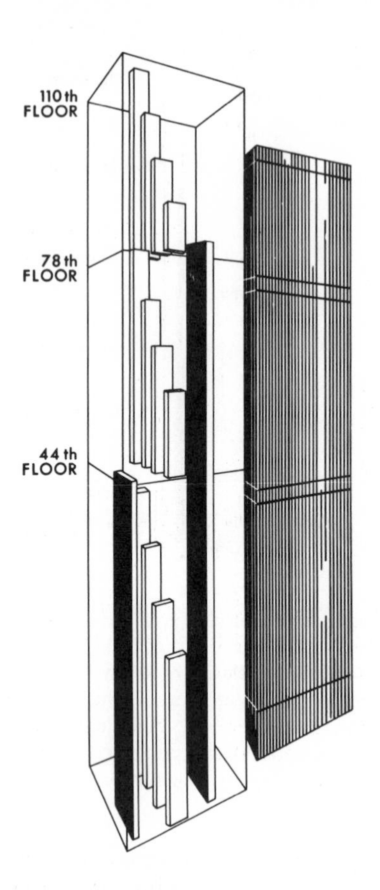

Fig. 10.10. Sky lobby elevator system for twin tower buildings of the World Trade Center in the Port of New York. Each 110-story tower has sky lobbies at the 44th and 78th floors. High-speed express elevators (solid) carry passengers nonstop to either sky lobby, where they take local elevators for their floors.

carrying capacity fall somewhat short of apparent gains because the plane or the elevator takes more time to reach its new, higher, top speed.

Taller buildings have brought faster elevators. Once the Empire State Building, in New York, had the world's fastest elevators, traveling at speeds up to 1,200 fpm, but the John Hancock Center now holds the record with some elevators traveling at 1,800 fpm.

Higher speed is coupled with improved automatic control to reduce total time from stop to stop, thereby reducing round-trip time and increasing elevator service. Modern elevators are designed, and their controls engineered, to minimize the time required for acceleration and retardation, door operation and passenger transfer.

Passenger comfort sets a ceiling on elevator speed and acceleration. People of varying age and physical condition are comfortable if elevators do not accelerate faster than $\frac{1}{8}$ g, equivalent to making a 160-lb man 20 lb "heavier," compared with a 7 to 9 g peak acceleration that astronauts may experience during lift-off.

Automatic operating systems for high-speed elevators accurately control acceleration and deceleration at rates that are consistent with passenger comfort, yet allow the car to travel at top speed for as much of the trip as possible. With powerful driving machines under precision control, elevators operate so smoothly that passengers can hardly feel their motion.

As a very high-rise, high-speed elevator rises or descends rapidly, change in atmospheric pressure from one altitude to another may be noticeable to some passengers. Pressurized elevator cars may be the solution to this problem.

10.9. *Sky Lobby Systems*

Hoistway space is effectively conserved by "sky lobby" elevator systems in very tall buildings. Such systems contribute materially to the economic feasibility of the John Hancock (Fig. 10.9) and World Trade Center (Fig. 10.10) towers.

Each tower is, in effect, two or three "buildings" stacked one over the other, with a sky lobby at the base of each "building" above the lowest. Passengers reach floors in the lowest "building" by local or express elevators in the usual way.

To reach floors in upper parts of the tower, passengers take high-speed, nonstop elevators directly to the sky lobby at the base of that section of the tower, where they change to other elevators for their floors. Hoistways of the latter elevators, from the sky lobby up to floors in the upper tower section, do not extend into the floors below and so overcome the primary problem of consuming lower-floor space.

Other factors also increase elevator service in proportion to the hoistway space used in these arrangements. Each of the elevators from the street level to the sky lobby levels and from there to the floors above does not travel as far as a conventional elevator going all the way from street level to upper floor.

Round-trip time is further reduced by nonstop operation from ground level to sky lobby. Sky lobby elevators have large, fast cars and achieve the highest possible handling capacity per square foot of hoistway.

Dining, shopping, and other facilities can be located at a sky lobby so that upper floor occupants need make fewer trips down to street level. Changing elevators at the sky lobby gives passengers an opportunity to adjust to the change in atmospheric pressure at a higher level.

An expanded version of the sky lobby system could make possible a "mile high" building, 400 stories tall. Conjecturally, ten 40-story buildings would be stacked one atop the other, with nine sky lobbies, at the 40th, 80th, 120th, 160th, 200th, 240th, 280th, 320th, and 360th floors. Nonstop elevators would connect each upper lobby with the street.

Conventionally arranged local and express elevators would serve floors up to the 39th. Other elevators operating up from the 40th floor would be similarly arranged in groups as for the first 39 floors of the building, comparable arrangements being made from each of the sky lobbies.

10.10. *Double-Deck Elevators*

Handling capacity can be increased by making elevator platforms larger to take more passengers per trip. But elevators with larger platforms need larger hoistways. To gain extra passenger capacity without enlarging hoistway dimensions an elevator may have two platforms, one above the other—the principle of the double-deck elevator (Fig. 10.11).

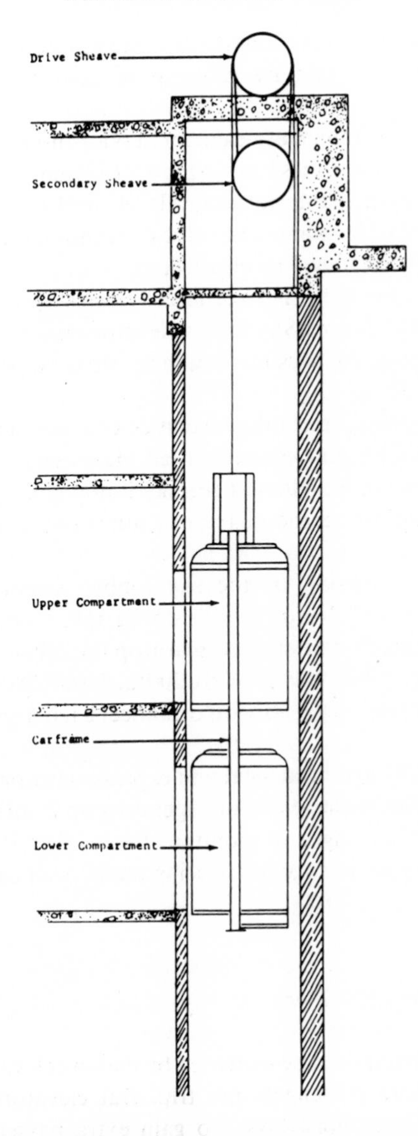

Fig. 10.11. Systems of double-deck elevators offer another means of providing ample service for very tall buildings without consuming too much space for hoistways. One compartment of each elevator would serve odd floors, the other, even floors.

An upper and a lower compartment are mounted in a car frame about twice the usual height. Passengers enter the compartments from two lower-terminal levels, possibly the street floor and a basement or mezzanine. People going to odd-numbered floors in the tower use one terminal level, to even-numbered floors, the other, possibly with escalators linking the two terminal levels.

As with other elevators controlled by automatic group supervisory systems, the double-deck elevator stops as it reaches floors for which passengers have touched buttons at landings or in either compartment. Some stops are made in response to calls by passengers in both compartments. At others, passengers enter or leave only one compartment while those in the other would wait momentarily.

Since each compartment is of normal size and carries the usual load, door operation and passenger entry and departure take no longer than in a conventional, single-deck elevator. At many stops the elevator will be serving, not one, but two floors. As a result the total number of stops per trip would be less than for a conventional elevator, further increasing the effective capacity of the double-deck installation.

For even greater carrying capacity the double-deck principle could well be applied to nonstop elevators in a sky lobby system with two-level sky lobbies. Upper and lower compartments of each double-deck elevator would serve, respectively, first floor and lower lobby at ground level and, at sky lobby levels, odd- and even-numbered floors. Escalators would connect lower and upper levels of each sky lobby.

10.11. *Two Cars in a Hoistway*

Space may also be saved by operating two separate elevators in a single hoistway (Fig. 10.12). As with a double-deck elevator, the two cars would have to start up from two different lower terminals, one at least a story above the other.

The upper elevator, an express, would leave first, followed at a safe distance by the lower car for local service. The lower, local car would make stops up to one floor below the first stop for the upper, express car. Going down, the local elevator would start first and reach its lower terminal before the express car reached its lower terminal.

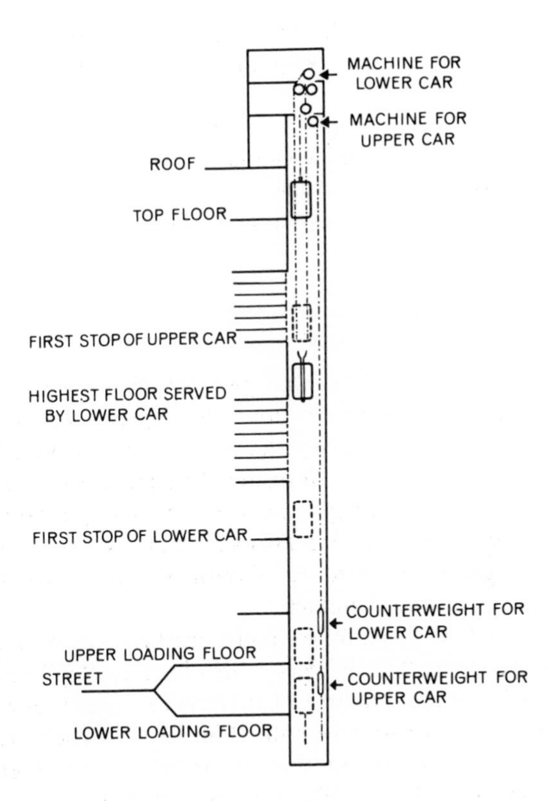

Fig. 10.12. Two or more elevators can be operated in a single hoistway to conserve space. Two cars would depart from a two-level main terminal, the upper car serving upper floors, the lower car, lower floors.

Each elevator would have its own driving machine and control system, but both cars would run on the same set of guide rails and both counterweights would similarly use the same counterweight rails. To guard against possible collision between the two cars, automatic controls for both would be electrically interlocked to stop either car before it came too close to the other.

Fail-safe mechanical protection would be provided by an oil buffer on a counterweight for interlocking the counterweights before both cars come too close. Scheduling between the two cars is critical and may, under certain traffic conditions, necessitate short extra waits for the upper or lower car.

The concept could be applied to three or even more cars using the same hoistway. The elevators would need three or more lower levels for loading or storage, and the spacing and timing of their operation would have to be controlled within very strict limits.

10.12. *Future Possibilities*

Sky lobby, double-deck, and two-cars-in-a-hoistway systems essentially apply present-day elevator technology. In the future, however, existing methods of moving elevators may be followed by other forms of propulsion, possibly eliminating hoisting ropes or supporting cylinders (Fig. 10.13). Such systems would permit unusual flexibility in operating many cars in one hoistway and also development of entirely new concepts in directing elevator movements to match the flow of traffic in a tower building.

During an "up" peak, for example, one hoistway would accommodate all the empty "down" cars, which for their "up" trips would be distributed among a number of hoistways to make stops at various floors in respónse to passenger calls (Fig. 10.14). When the flow of traffic reverses, operation of the system would be reversed. As elevators complete trips, horizontal transfer mechanisms would feed them from the active to the return hoistway, or vice versa.

In the active hoistways, elevators would follow one another at intervals controlled to minimize waiting time and maximize handling capacity. Some of the active hoistways would accommodate "up" cars while others were used by "down" cars, the service varying in proportion to the changing pattern of two-way traffic.

The self-propelled elevator system resembles an automated mass transit system set on end. Transfer arrangements at intermediate floors, like switches on the transit line, would facilitate control of elevator operation to make full use of hoistway capacity and provide the most expeditious service. Sensing devices would detect empty cars, which could be fed via intermediate transfer levels into portions of the building where demand was heaviest.

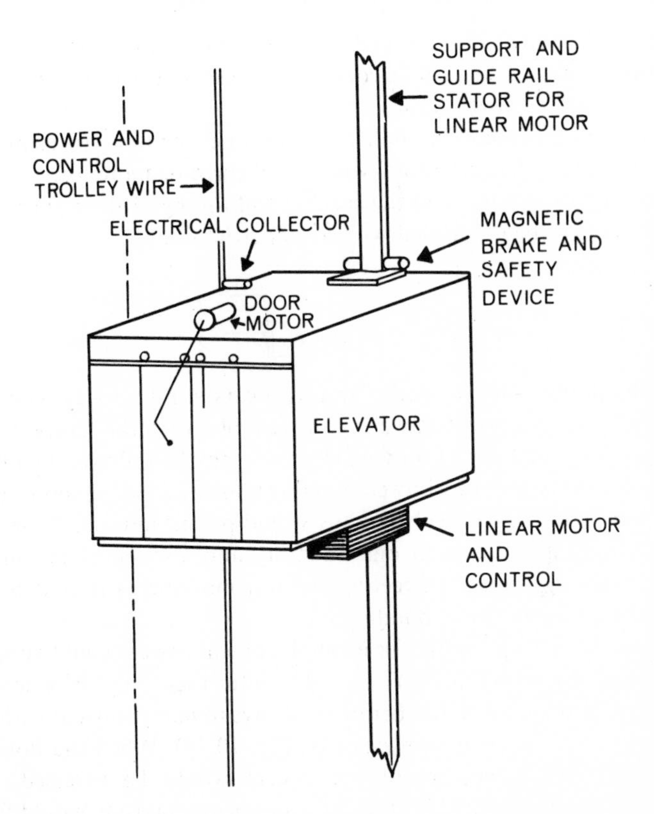

Fig. 10.13. Future elevators may be self-propelled, permitting great flexibility in operating several cars in a single hoistway. "Linear induction motors" or other forms of propulsion could replace conventional rope and traction machine systems.

10.13. *In-Building Parking*

As the use of private automobiles has increased, the need for off-street parking facilities has become urgent, especially in central business districts with clusters of high-rise buildings. Otherwise many persons are unable to find parking space near their destinations, while those who do park their cars in the street deny moving traffic use of at least one or two lanes.

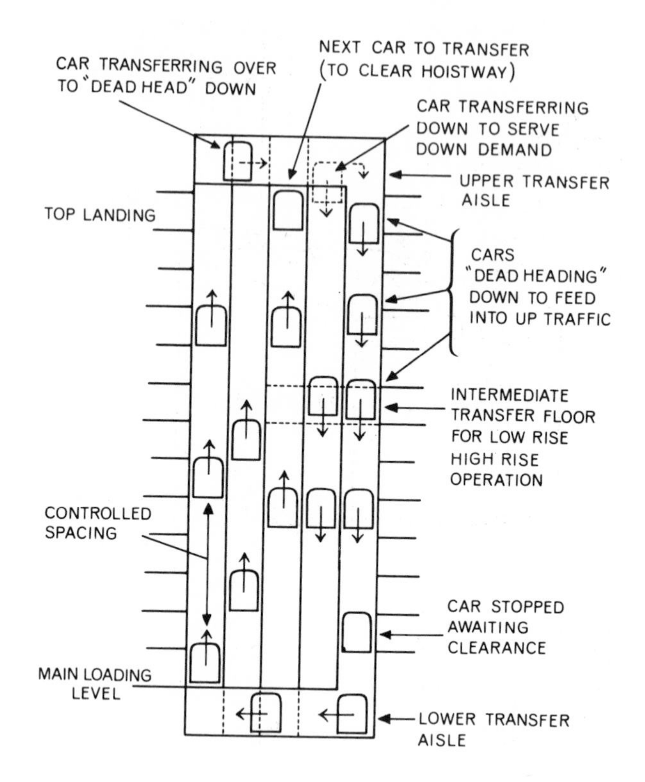

Fig. 10.14. Self-propelled elevators could be spaced in separate hoistways for fast, frequent service in the direction of heavier traffic, then, as available, shunted to a common hoistway for return to the load center.

Downtown land costs usually necessitate multilevel parking, either in separate garages or in garages integrated with commercial or residential buildings. Convenience of in-building parking to tenants and visitors makes it a decided economic asset to the owner of the building.

Automobiles may be parked on upper levels by being carried on a freight elevator or driven up a series of ramps. In an elevator facility of the conventional type, cars must be parked by attendants. But a ramp garage may be built for either attendant or customer parking.

10.14. *Automated Parking Systems*

A computer-controlled system has been developed especially for the purpose of using otherwise marginal space in a building's basement or core to park as many as several hundred cars (Fig. 10.15).

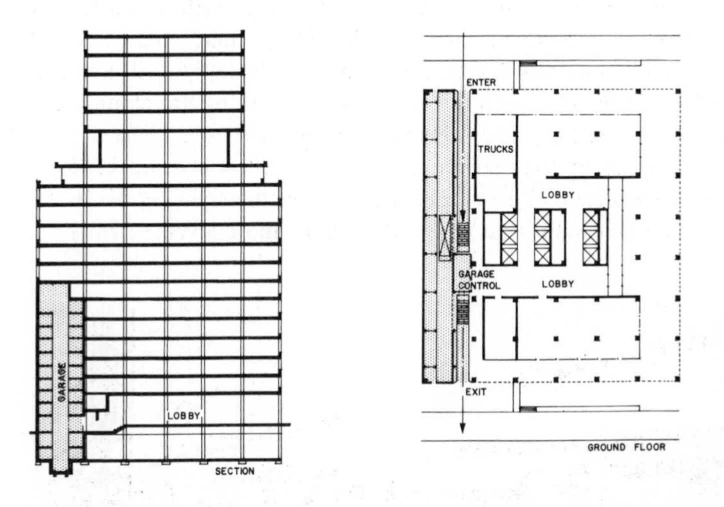

Fig. 10.15. An automated parking system occupies narrow marginal space in a building. Computer-controlled tower-elevator moves automobiles horizontally and vertically to park them in compactly arranged stalls. Cars always face in the same direction, need no space for turning or maneuvering.

An integral part of the building, the system parks each car for a few minutes or many hours and promptly returns it to the lobby, ready to be driven off again.

Within the automated parking facility, cars are not driven but are handled mechanically and always face in the same direction, saving space otherwise needed for maneuvering or turning, or for ramps from level to level. Cars can be parked on many more levels than in a ramp facility, and the height per level is less. The automated system not only uses less volume per car parked, but also the space may be relatively narrow and located almost anywhere in the building.

Less cubage per car reduces building construction or excavation costs. Since car engines do not run within the automated facility, it needs no high-capacity ventilating system for exhaust fumes.

An automated facility for transient parkers is controlled by an electronic computer that also calculates parking charges according to any desired schedules, with a single cashier-attendant in charge of the entire facility. In an apartment building, with all parking stalls permanently assigned, tenants themselves may initiate the fully automated operation. Either way, most of the labor costs and personnel problems of conventional, attendant parking are eliminated.

The heart of the automated parking system is a special moving-tower elevator that carried cars both horizontally and vertically to rows of stalls on either side of the elevator runways (Fig. 10.16). There may be nine stalls in a row and seven or more levels or rows.

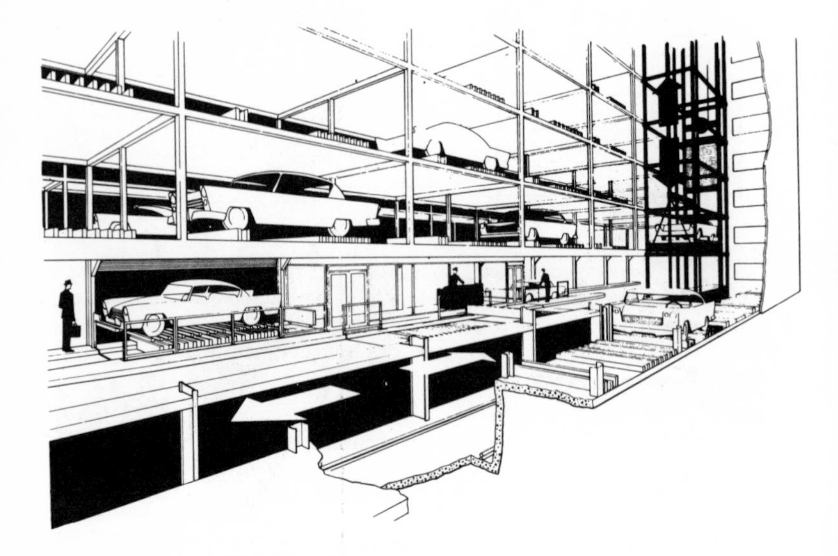

Fig. 10.16. Cars are parked completely automatically in stalls on either side of tower-elevator runway. Special fingers reach under tire treads to lift cars on and off elevator.

A facility with a single tower elevator and spaces for as many as 151 cars can be installed in a section of a building 24 ft. wide by 200 ft. long. An installation with two elevators and stalls for 276 cars would require a space 48 ft. wide. Space for elevator runways and parking stalls can be in the basement or above grade.

Where space is limited longitudinally, the system may use elevators which move vertically only, parking two cars on each level, on opposite sides of the hoistway. Such an arrangement can park cars on as many as 64 levels, equivalent to about 40 office floors. With two elevators placed back-to-back, the "tower within a tower" could park 248 cars on a ground area only 25 by 50 ft.

Whether the building's parking core is of longitudinal or tower configuration, using space above grade, the computer console that controls the entire facility may be located at street level near the parking and unparking stations and the entrance to the passenger-elevator lobby (Fig. 10.17).

Fig. 10.17. The parking facility is controlled completely automatically by a computer console, which may be in the building lobby. One person, a cashier-attendant, can easily operate the entire system.

As users gain experience with systems like this, automated in-building parking will become a service that they take for granted in centrally located buildings, as they now accept and expect central air conditioning and modern vertical transportation.

10.15. *Multipurpose Buildings*

John Hancock Center is, in effect, a 48-floor apartment tower atop 28 floors of offices atop a parking garage, stores, and services. It is one of a growing number of buildings constructed or proposed in recent years to house two or more different kinds of facilities (Table 10.3) under one roof.

Table 10.3
Typical facilities combined in multipurpose buildings

Apartment and elementary school	Hospital and nurses' residence
Apartment and office	Office and light manufacturing
College classroom and residence	Offices, retail, restaurant
Department store and hotel	Post office and business office
Exhibit and office	Parking garage with any of above

One reason why multi-use buildings are now attracting more attention than ever before is the long-lasting building boom that has gobbled up desirable in-town sites and pushed up the cost of those that remain.

More intensive use of land as well as of building structure and its essential services also contributes to the appeal of the combination building. When people live as well as work in a building, its earning power continues around the clock and throughout the calendar.

Two-in-one design offers other advantages. A variety of different services under the same roof increases a building's convenience for tenants and its attraction for visitors. The possibility of commuting from home to office simply by riding an elevator can be a blessing in an era of ever greater congestion in horizontal travel.

Successful multipurpose design depends largely on properly locating the different functions in the building and providing adequate vertical transportation to its various levels. Vertical separation, with no two major functions combined on a single floor, usually works best. Offices, for instance, should be on certain floors of a tower, apartments on others. A mechanical floor or a sky lobby may separate residential from commercial floors.

With separation, living accommodations can occupy the higher floors, which remain more desirable for the purpose despite progress

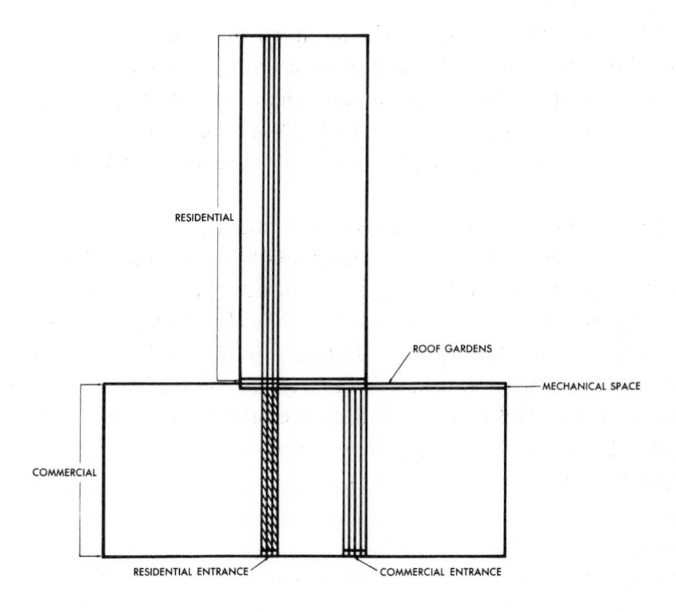

Fig. 10.18. Vertical separation of functions in a multipurpose building. To separate traffic streams, each portion of the building has its own entrance lobby and elevator system.

in air conditioning and artificial lighting. Safety of children and security of all residents are best assured by locating them on levels distinct from those used for commercial purposes.

Traffic streams to different parts of the building should be kept from interfering with each other (Fig. 10.18). In a downtown apartment-office tower, for example, some residents may also work in the same building and expect convenient transportation between both sections of the building.

But many if not most of the apartment tenants may work elsewhere, while most office workers and callers come from outside the building. Traffic could conflict in the morning, when people are leaving apartments in the building to go to work while others are arriving en route to their offices, and interference might recur at the end of the working day.

In an office building with an upper-floor restaurant, vertical traffic flows are also likely to oppose each other. Office tenants will be going down to lunch or dinner when restaurant patrons want service up to that facility. At the end of the lunch period, traffic from the restaurant will be mostly down, but predominantly up to the office floors.

Certain kinds of traffic should never mingle. For example, children living in the apartment portion of a combination building should be kept from playing in the office lobby or riding in the office elevators.

A hotel or motel with extensive facilities for public functions may be considered a multipurpose building. Traffic to ballrooms, meeting rooms, and restaurants should be separated from traffic to guest room floors. Guests, even those attending a convention or other affair in the hotel, appreciate quiet, uncongested elevator service to their sleeping quarters.

10.16. *Separate or Shared Transportation*

Conditions like these, often occurring in downtown multifunction buildings, usually require that each function have its own entrance lobby and its own elevators or escalators. In that way the two (or more) streams of traffic will be kept from interfering with each other.

Each part of the combination building may be treated as an independent building in determining the number, size, and speed of elevators or escalators it needs. They should provide sufficient vertical transportation to accommodate the heaviest periods of traffic expected to the floors served. Transportation to each part of the building should also be prompt enough to satisfy the respective demands, for example, of apartment and office occupants.

Lower floors devoted to a department or specialty store may be served primarily by escalators, with high-speed elevators to apartment, hotel, or office floors above. In other cases, levels devoted to different purposes can have their own groups of low-rise or high-rise elevators, much like local and express groups in conventional buildings.

A sky lobby arrangement achieves the goal of separating diverse streams of traffic while transporting passengers to their floors quickly, even high up in the tower. For years Cincinnati's Terrace Hilton Hotel has been successfully using such an arrangement. The hotel proper is over a store with its own elevators and escalators. Separate elevators take hotel guests to a lobby at the terrace level, above the store, and from there up to guest room floors.

Sometimes, as in a combination college classroom-residence building or elementary school-apartment building, both functions may share a common group of elevators. Sharing elevators is usually practical only when many of the same people use both portions of the building. During much of the day, when traffic to classroom floors is heavy, traffic to apartment or dormitory floors will be light, and vice versa. With limited conflict between streams of traffic, one elevator group may serve both.

The elevator (or escalator) system should have sufficient carrying capacity for the traffic generated by the busier of such a building's two functions. Capacity will also be ample for the traffic of the building's less active function, as long as traffic to both parts of the building reaches a maximum at different times.

Where elevators are shared by passengers to both classroom and residential floors, special provisions may be incorporated for security of the latter. Automatic elevators may be equipped with controls operated by keyswitches so that cars stop at apartment or residence floors only for calls from authorized persons using the appropriate keys.

Multipurpose buildings of various sizes and shapes represent a promising response to present and prospective demands. Vertical transportation engineered for each such building will contribute to the economy and effectiveness with which it serves its owners and users.

Chapter 11

Appearance, Comfort, and Convenience

To the architect, elevators and escalators have become more than a means of transportation; they constitute an integral element of a building's interior—and, in some cases, exterior—design (Fig. 11.1). With increasing perfection in equipment operation, users are more decisively impressed by appearance and related qualities.

Elevator automation has given visual factors a more active functional as well as a purely esthetic role. Without elevator atten-

Fig. 11.1. Elevators and escalators contribute to architectural distinction of lobbies in Chase Manhattan Bank Building, New York. Skidmore, Owings & Merrill, architect-engineer; Turner Construction Co, general contractor.

dants to call floors and operate controls, passengers must see with unmistakable clarity which car to enter, which button to touch, and at which stop to leave. Passengers even seem to enter an operatorless car more readily if it is well lighted and inviting in appearance.

Responding to the challenge, architects are achieving notable results in using new elevator and escalator designs and materials to add esthetic appeal. Passenger satisfaction is also furthered by electronic communication systems that provide continuous contact with the building staff.

Application of these developments is not limited to new buildings. Many elevator interiors and entrances embodying advanced design concepts are the result of modernization projects which retain much of the original working equipment. New methods are also making it economical and practical to bring existing escalators up to contemporary standards of appearance as well as performance.

11.1. *Car Design and Finish*

While every element of the elevator installation performs an essential function, the car and its platform are the portions of the system that directly serve passengers riding from floor to floor. Car and platform are mounted on the car frame which moves up and down in the hoistway.

Of this equipment, passengers normally see little more than the interior of the car or cab. In design and finish, cars should harmonize with the building in which they are installed, especially its elevator lobby and corridors.

Modern car interiors often owe their visual interest to natural metallic finishes, including extruded aluminum, bronze, and nickel-silver and rigidized stainless steel. Various surface treatments may be applied to these metals to achieve desired color and design effects.

Increasingly popular interiors combine a natural metal front with replaceable side and rear panels of wood faced with veneer or high-pressure plastic laminates (Fig. 11.2). Base, trim, and hand rails are also often finished in natural metal. Mounted in the all-steel shell, the side panels can be removed and replaced without taking the car out of service.

Fig. 11.2. Modern elevator cab treatment is exemplified by this design incorporating side panels of wood faced with high-pressure plastic laminates and stainless steel front.

From the replaceable panel evolved the reversible panel, both sides of which have finished surfaces. Side and rear wall panels may readily be lifted and turned to change a car interior in minutes (Figs. 11.3 to 11.6).

Opposite surfaces of each panel are finished in different colors, designs, or materials. Not only can panels be arranged to create varied interior designs but also, because either side may be exposed as a finished surface, reversible panels wear twice as long as the usual type.

11.2. *Car Interiors*

Elevators may have interiors of an individualized design characteristic of the particular building in which they are installed. Instead of similar treatment for all three walls of a car, its sides and rear may differ in design and materials. Besides veneers and laminates, walls have also been faced with metal extrusions, oil-finished natural hardwood, and even marble.

Distinctive murals may be created on car walls by the use of colored ceramic tiles (Fig. 11.7), photographic blowups, or other techniques. Decorative murals may be "painted" on stainless steel wall panels by working the surface of the metal with abrasives.

Striking effects in elevator cars have also been achieved with luminous rear walls. Such cars have wall panels of polyester fiber glass, cast to integrate colored design with textured relief. Illumination is from the rear with fluorescent tubes. Wall panels may be plain or may incorporate abstract or decorative designs.

11.3. *Lighting the Car*

In modern elevator cars, lighting is integrated with other elements of interior design. As has been mentioned, high-level, glare-free illumination can actually speed elevator service because people will walk into an operatorless car without hesitation if it is at least as bright as or brighter than the elevator lobby or hall.

Lighting methods and ceiling designs are usually closely related. Uniform illumination over the entire ceiling area is achieved by

Figs. 11.3–11.6. These four renderings all show the same car with reversible side and rear panels. Reversible panels have differently finished surfaces on both sides, can be lifted and turned to change car interiors.

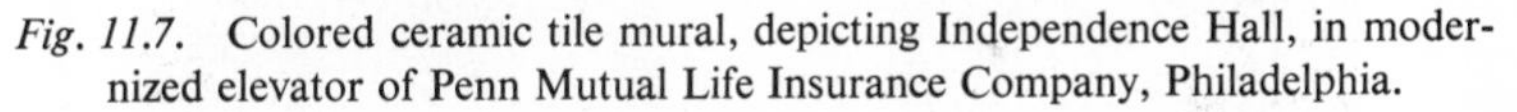

Fig. 11.7. Colored ceramic tile mural, depicting Independence Hall, in modernized elevator of Penn Mutual Life Insurance Company, Philadelphia.

fluorescent lamps above a suspended ceiling of eggcrate grilles (Fig. 11.8) or acrylic panels.

Patterns of concentrated lighting create dramatic effects. Incandescent lamps above cutouts in a suspended ceiling serve as down spots lighting the front of the car and threshold, supplemented by fluorescent lamps above grilles across the back of the car or around sides and back (Fig. 11.9).

Down spots may provide all the illumination for a cab. Or the suspended ceiling may be pierced with clear acrylic rods or prisms

Fig. 11.8. Car is uniformly illuminated by fluorescent lamps above suspended eggcrate ceiling. Exhaust blower above suspended ceiling provides ventilation without drafts.

conducting light from fluorescent lamps above the ceiling and creating a jeweled effect in contrast with the dark-finished ceiling (Fig. 11.10).

Fig. 11.9. Grilles around three sides provide direct lighting from fluorescent lamps above hung ceiling, with down lights (incandescent) across entrance. Walls have natural metal dado below plastic laminate wood panels.

11.4. *Exhaust and Pressure Ventilation*

So that people in today's fully enclosed elevator cars will feel comfortable, even on hot, humid days, stale air in the car should be continuously replaced by fresh. During periods of peak traffic the

Fig. 11.10. Direct fluorescent lighting through 24 clear acrylic rods in suspended ceilings. Car operating fixtures integral with front return panels. Front and side panels are formica, other surfaces aluminum.

ventilating system should completely change the air in the car twice every minute. Whether or not the car is moving, passengers should be able to feel the air in motion, yet not be subjected to uncomfortable drafts.

Elevator cars are ventilated by exhaust or pressure blowers. In an air-conditioned building, exhaust ventilation is used to draw air into a car through its open doors when the car is stopped at a floor. Air is exhausted from the car through a ceiling grille or around the perimeter of a suspended ceiling and discharged by the blower into the hoistway.

In a building without air conditioning, a pressure blower atop the elevator forces hoistway air into the cab directly through a spreader grille in the ceiling of a small cab. In a larger cab, uniform, draftless diffusion of air into the passenger enclosure requires a plenum chamber into which the blower forces air. From the plenum, air enters all parts of the cab at equal pressure through perforations around the ceiling perimeter.

11.5. *Sound Isolation*

People prefer elevators that are not only comfortably ventilated but also operate with minimum noise or vibration. Measures to control objectionable sound originating in machine rooms and hoistways are outlined in Sections 4.10 and 4.11. For maximum riding comfort, modern practice also prescribes sound isolation between the car and its supporting frame.

Isolation is achieved by having the entire car, including its platform, rest on resilient rubber blocks mounted on the steel supporting structure (Fig. 11.11). Rubber-faced clamps welded to each side of the car canopy firmly fasten the top of the car to the car frame upright members. Eliminating metal-to-metal contact between car and frame blocks sound transmission, dampens vibration, and cushions car movement.

For further isolation, the car frame moves on rubber-tired roller guides along the hoistway guide rails, car doors are suspended on rubber-tired hangers, and car ventilating blowers are mounted on rubber pads. Metal cars have side panels coated on the outside with

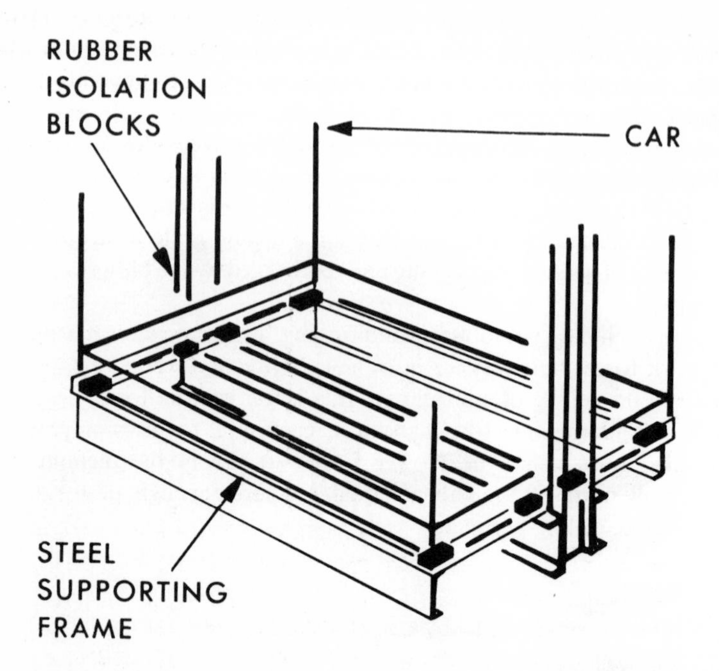

Fig. 11.11. For sound isolation, the entire elevator car, including its platform, rests on rubber blocks on the steel supporting frame, avoiding metal-to-metal contact between car and car frame. Locking blocks restrain lateral and upward motion of the car.

a sound-deadening compound that absorbs the noise of passing air and prevents amplification of sounds from within the car.

11.6. *Elevators in the Open*

Although conventional elevators display only their entrances and car interiors, dramatic exceptions are found in the growing number of transparent-walled observation elevators designed for a high level of visibility. Not only can people outside see the elevator and its passengers, but also the latter enjoy a "ride with a view" which can be a stimulating experience.

A notable installation of this type serves the Regency Hyatt House in Atlanta, Georgia. In the glass-roofed "atrium," 23 stories high, that distinguishes the hotel architecturally a group of five high-speed elevators operates in full view. The cars have side panels of safety glass and in appearance somewhat suggest space vehicles (Fig. 11.12).

Fig. 11.12. One of five glass-walled observation elevators in central court of Regency Hyatt House, Atlanta, Georgia. Edwards and Portman, architect and engineer; J. A. Jones Construction Company, general contractor.

Traveling on the outside of an elevator tower projecting into the central court, the elevators form an integral visual element of the building. From the lobby below and the balcony floors above, the cars gliding up and down appear like a giant, sculptural mobile, while passengers in the transparent elevators enjoy changing views as they rise through the court.

Elevators like these, operating in the open, often have cylindrically shaped cars with tapered tops and bottoms and seem to move almost without visible means of support. Operating mechanisms on top of the usual elevator are concealed within domes atop these cars. Car and counterweight ropes may be located close to the building wall, where they are relatively inconspicuous.

Towers or buildings commanding scenic views have elevators of this type to let passengers see the spectacle as they ride. Some of these installations are in special structures at major expositions, like the permanent 600-ft. Space Needle in Seattle, Washington, and the still taller Tower of the Americas at San Antonio, Texas.

In resort and commercial hotels and office skyscrapers as well, observation elevators add an extra dimension of enjoyment to practical transportation (Fig. 11.13). Many of these buildings use the glass-walled cars to carry patrons to restaurants or observation galleries at commanding heights and make the scenic vertical ride part of a uniquely memorable experience.

11.7. *Elevator Entrances*

Whether elevators operate in full view or in conventional hoistways, they need suitable entrances at every landing. Structurally, an elevator entrance includes the frame, door panels, sill, header, and other parts (Fig. 11.14). Doors of the single-speed, two-speed, and center-opening types are accommodated by appropriate entrance arrangements.

Functionally as well as esthetically, entrances should be planned to serve as a transition between the elevator lobby or hall and the car interior. In design and materials, consequently, entrances usually harmonize with elevator cars as well as with adjacent building spaces.

Entrance frames and doors are heavy-gauge steel. Upper-floor

Fig. 11.13. Supplementing eight conventional elevators in the 31-story Security Life Building in Denver, Colorado, this glass-walled elevator travels in a glass-enclosed shaft to a restaurant overlooking the scenic city and surrounding Colorado Rockies. Sorey, Hill and Sorey, architect and engineer; Harmon Construction Company, general contractor.

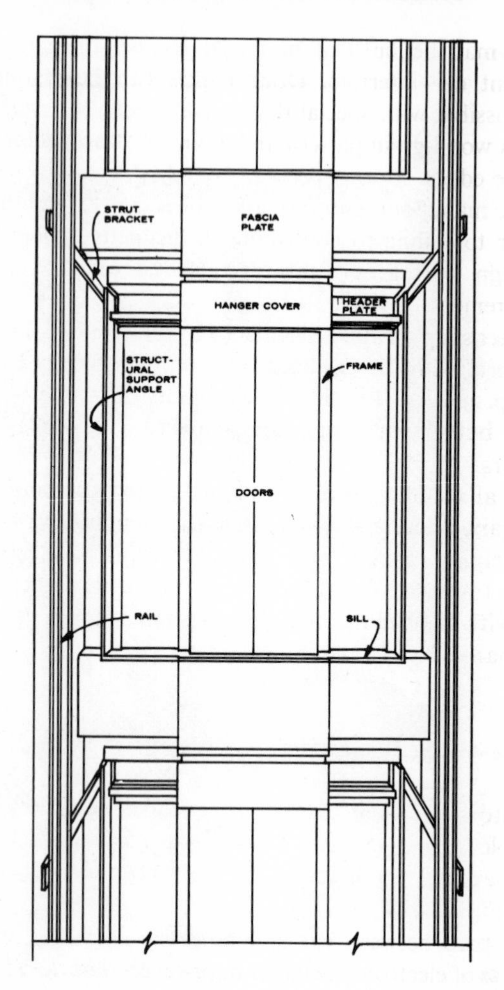

Fig. 11.14. Principal structural elements of an elevator entrance, from the hoistway side.

entrances may be finished in baked enamel or natural metal to complement car interiors. Door panels can be stainless steel or bronze, possibly with special decorative effects, or laminated plastic finish with wood-grain pattern and a binder of bronze or other metal around the edges.

At the main floor, entrance assemblies may extend the full height from floor to ceiling to contribute to distinctive effects in elevator lobby design. Entrances of this type also reduce lobby wall construction requirements.

Further simplification is achieved by directional lanterns (Section 11.9) integral with the entrance transom, which may be flush (Fig. 11.15) or projecting (Fig. 11.16). Entrances shown have center-opening doors, but similar designs are also used with single-slide or two-speed doors.

Natural metal finishes are often favored for entrance frames, transoms, and door panels, and extruded aluminum or bronze for sills. Alternatively, aluminum or bronze extrusions have been used for doors and transoms. Other treatments for entrance framing include finishing with anodized aluminum and marble jambs to match lobby walls of marble.

11.8. *Close-Coupled Entrances*

Elevators and their entrances are still more fully integrated by close-coupled arrangements. These designs gain esthetic and functional advantages by bringing car and hoistway entrances closer together (Fig. 11.17).

Less space between car and hoistway entrances improves the effectiveness of electronic-detector door-reversal devices for passenger protection (Section 5.4). More closely integrated car and entrance design reduces the depth of the "tunnel effect" created by usual arrangements.

Narrower car and hoistway sills of the close-coupled assemblies result in a shallower, more pleasing entrance. Visually, the entire car and entrance become more intimately associated with the elevator corridor or lobby.

The shallower entrance arrangement saves space to make the

Fig. 11.15. Elevator entrance extending all the way from floor to ceiling gains appearance of slender height. In this design, with flush transom, mock astragal at center of transom simulates door astragal.

elevator roomier without enlarging the hoistway. Alternatively, hoistway depth can be reduced, releasing core area for wider corridors or other purposes.

Fig. 11.16. Full-height entrance with projecting transom. Directional lanterns integral with transom achieve further simplification of design and construction. Layouts must be so planned that all passengers waiting in the elevator corridor or lobby can see the lanterns.

11.9. *Entrance Safety Devices*

Because most elevator accidents occur at elevator entrances, provisions for passenger safety are an essential part of every entrance. Protection is assured by devices that prevent an entrance from opening unless the car has come to a full stop at the landing, and prevent a car from moving unless all entrances, car and hoistway, are safely closed.

Hoistway entrances have electrical-mechanical interlocks that prevent opening of doors if the car is not at the landing, and movement of the car if any hoistway door is open. Electrical contacts on car doors prevent movement of the car unless its doors are closed all the way.

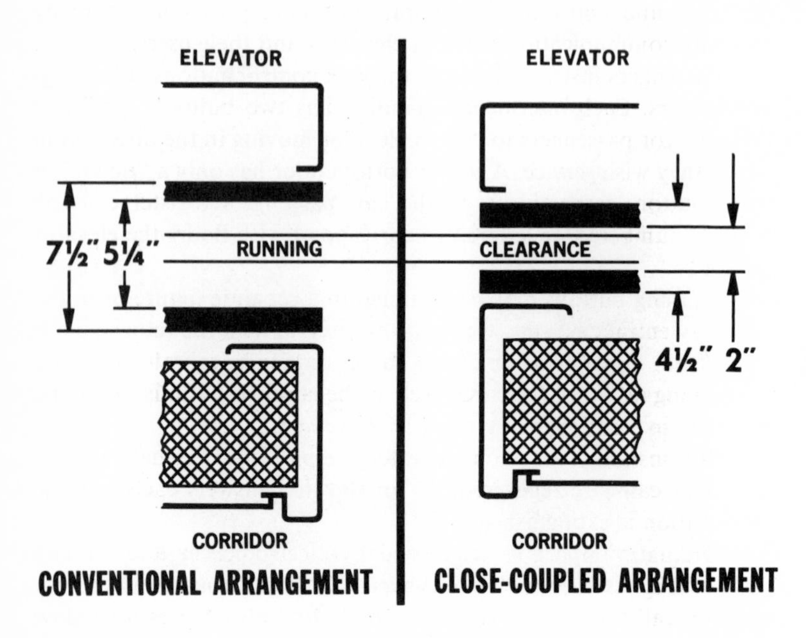

Fig. 11.17. Comparative plans indicate how close-coupled arrangements (right) bring elevator car and hoistway entrances closer together to gain esthetic and functional advantages.

Automatic elevators, require, in addition, door-reversal devices for passenger convenience and protection. Pressure-actuated safety shoes, light beam-photocell systems, and electronic proximity detectors working separately or in combination automatically stop and reverse closing doors if a passenger is in the way.

11.10. *Control and Communication*

Even in manually operated elevators, attendants need at least simple controls to direct the movement and stopping of the elevator and, usually, indicators to show the position of the car. Since modern, automatic elevators usually have no attendants, it is still more urgent that cars and entrances incorporate adequate provision for ready two-way communication between elevators and their users.

Passengers instruct elevators through control buttons at landings and in cars. Each intermediate landing has two buttons, "up" and "down," for passengers to call an elevator moving in the direction in which they wish service. A top or bottom floor has only a "down" or "up" button, respectively. In the car, passengers register calls on buttons numbered or lettered to correspond with floors the elevator serves.

Landing buttons may be mounted in a separate fixture or in the hoistway entrance frame. Car buttons integral with the return panels at the front of the car (Fig. 11.18), rather than in a separate faceplate, are gaining favor. In the car as well as the landing, buttons should be mounted in locations convenient to passengers.

Buttons are often illuminated to give passengers visual evidence that their calls are registered. As an elevator answers each call, the illumination is extinguished.

Dramatic impact is heightened by electronic car and landing buttons (Fig. 11.19) which passengers need merely touch, rather than push, to call for service. Touching the button also causes it to glow and to remain illuminated until the call is answered. Buttons are designed to respond only to a positive touch and not to persons accidentally brushing against them.

11.11. *Car and Hallway Signals*

Communicating essential information from elevators to passengers is usually accomplished visually, by signals in cars and hallways, but may also be achieved by audible means. Indicators, lanterns, and other signals must conform esthetically with the car interior or corridor designs of which they form a part, but their functional role is primary.

Alerting passengers to elevator arrivals and departures, signals encourage promptness in entering and leaving elevators and thereby speed service for all users. With the trend to elevator automation, signals have been redesigned to make their numerals, letters, or other indications more easily legible to passengers.

Fig. 11.18. Car operating buttons integral with front panels, below digital readout car position indicators.

In the car the usual signal is a car position indicator to show passengers the direction in which the elevator is moving and the floor at which it is stopping. At landings a similar indicator may be mounted over each entrance (Fig. 11.20), although separate directional lanterns are usually preferred.

Fig. 11.19. Landing fixtures for elevator corridors may incorporate electronic touch buttons (left) or mechanical push buttons (right), with floor buttons of corresponding type in the car. Electronic buttons glow when touched to register calls, are extinguished when elevator answers.

Directional lanterns may be integral with the elevator entrance header (Fig. 11.16) or side jamb (Fig. 11.21) or mounted in the corridor wall (Fig. 11.15). As the lantern lights, a chime or gong may also sound to attract passengers' attention to an arriving elevator.

For elevators serving many stops, conventional car position

indicators, with numeral cutouts for every floor, may take too much space. For more compact, pleasing designs, digital readouts like those on electronic computers are being used as floor indicators in elevator cars (Fig. 11.18), corridors, and lobby panels (Fig. 11.22).

A high-intensity lamp projects images of standard numerals or

Fig. 11.20. Car position indicator with directional lantern over elevator entrance.

Fig. 11.21. Directional lanterns in side jambs of elevator entrance frames.

letters onto a viewing screen, one screen sufficing for all floors. Whereas conventional lobby panels for high-rise buildings may be 6 or 7 ft. high, a readout panel requires only 15 in. to display the same information. Readout indicators in the car permit comparable improvement in design.

11.12. *Audible Communication Systems*

When elevators were manually operated, telephones were often installed in elevator cars to link attendants with the starter or other building personnel. In an emergency the attendant could telephone to his superiors to summon help.

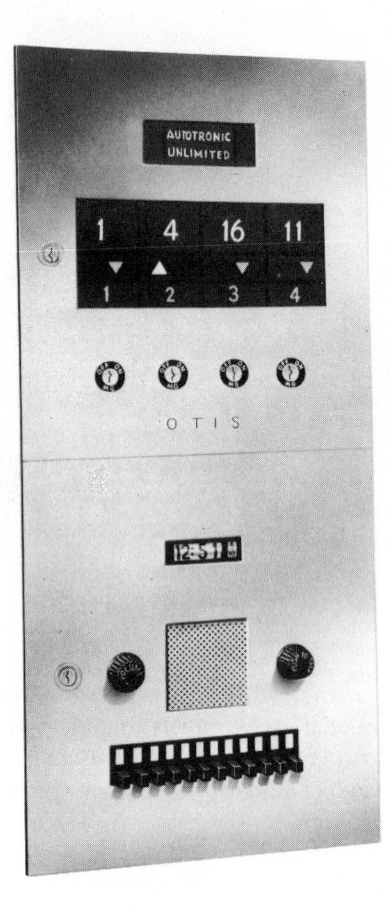

Fig. 11.22. Lobby panel with digital readout car position indicator in upper portion. This part of the panel is only 15 in. high, instead of the 6 or 7 ft. that would be required for a conventional-type indicator in a high-rise building.

Conventional telephones may be retained in automatic elevators for use directly by passengers. But a more appropriate form of two-way voice communication developed specifically for unattended elevators affords passengers the secure feeling of continuous contact with building personnel.

Instead of the usual telephone instruments, each car has an intercom system. Loudspeakers in the cars are connected to similar equipment at the starter's station, building engineer's office, or other location. Circuits are so designed that the starter or other authorized member of the building staff can listen or speak to passengers in any selected elevator or elevators.

11.13. *Emergency Power Systems*

As a precaution against possible failure in the supply of electricity to a building, emergency power systems assure at least a minimum of energy for elevator lighting, communication, or operation. Contributing importantly to safety, such systems also enable building personnel to reassure passengers in automatic elevators stalled by power failure.

Compared with the value of emergency power for elevator lighting and communication in making passengers feel easier and speeding efforts to release them from an elevator caught between floors, the cost of such arrangements is nominal. Essentially, the power source for this purpose is a charger-storage battery-converter combination to supply alternating current for fluorescent or incandescent lamps and direct current for telephone or speaker-type intercom equipment.

Automatic controls keep the battery fully charged and switch it to recharge once normal power is restored. Should outside power fail, the emergency system automatically starts its converter and assumes the lighting and communication electrical load.

If a building has generators driven by diesel or gasoline engines or gas turbines, they may supply power for elevators. Electric power from the building's generating plant assures at least partial operation of elevator and other essential services during an emergency.

To minimize loads on the emergency generator, elevators may be operated one car at a time, possibly at reduced speed. In a building with several elevators, the cars which are to use emergency power can be selected manually or automatically.

Manual selection may be made by a key-operated switch in the building lobby. This arrangement is more economical to install but depends on a member of the building staff being available to operate the manual selector switch.

With automatic selection, cars start automatically on emergency power, one at a time, and move to the main floor. One elevator then remains in service on power from the standby generator. If elevators are the electric-traction type with variable voltage control, they can operate at a reduced speed to lessen demand on the emergency power generator.

For a hydraulic elevator, emergency electricity can be supplied to the valve solenoids and door operator even if power is insufficient to drive the pump motor. The car then descends safely to a lower floor and releases its passengers. This automatic arrangement is in addition

Fig. 11.23. Transparent-balustrade escalators offer new opportunities for visual integration of vertical transportation system with building interiors and exteriors. Transparent balustrades make areas appear more spacious.

to the manually operated valve for lowering a hydraulic elevator in an emergency.

11.14. *Escalators as Design Elements*

Perhaps even more than elevators, escalators have been treated effectively as elements of building design, possibly because their form resembles the stairs long familiar to architects. In commercial and institutional buildings around the world, escalators not only provide efficient vertical transportation but also contribute to distinctive architectural effects in lobbies and on upper floors.

Fig. 11.24. Exceptionally slender silhouette of escalators in General Motors Building, New York, is achieved with trusses less than half as deep as usual. The building structure incorporates special provisions for the moving steps to return behind the wall and floor rather than with truss. Edward Durell Stone and Emery Roth and Sons, associate architects; George A. Fuller Company, general contractor.

Escalator balustrades have finishes and trim like those of elevator cars and entrances. Balustrade panels may be finished in baked enamel in any of a variety of colors, or in natural metals treated in various ways, often with the metal finishes for decks and skirting.

Escalators with balustrades of transparent safety glass are gaining favor. Blending into their background, "see-through" escalators make building areas seem more open and spacious (Fig. 11.23).

Transparent-balustrade escalators facilitate imaginative architectural treatments that dramatize building interiors and exteriors. To create desired visual effects, balustrades may be glass or plastic, clear, translucent, tinted, textured, or with special designs.

Innovations in escalators design promise to permit a truss less than half as deep as usual, resulting in outlines so slender that riders seem to be rising through space (Fig. 11.24). In such installations the building structure must incorporate provisions for the escalator steps to return behind and beneath an adjacent wall and floor rather than within the truss, the usual arrangement.

Elevators and escalators offer the architect a growing range of esthetic options. As he acquires greater understanding of these possibilities and their use, the building designer increases his own power to handle creatively the visual challenge of vertical transportation.

Chapter 12

Modernizing Vertical Transportation

Booming construction in recent years has brought a multitude of modern buildings in cities around the world. Most of these buildings boast advanced design and highly efficient services, including vertical transportation.

Higher standards established by the growing number of new buildings tend to raise the levels of service expected by tenants and users of all buildings. Older structures, finding competition increasingly intense, face declines in occupancy and rentals.

Yet older buildings, if strategically located, soundly constructed, and of basically suitable design, can be modernized to compete economically with newer structures. Architects and consulting engineers are often asked to plan programs for modernizing electrical, mechanical, and vertical transportation facilities in such buildings.

Elevators and escalators installed decades ago may be economically brought up to present-day standards of performance and appearance by retaining and reusing much of the original machinery (Fig. 12.1). Elimination of attendants through elevator modernization may reduce operating costs sufficiently to amortize the improvement program over a period of years.

12.1. *Elevator Modernization Economics*

Converting elevators from manual to automatic operation brings gains in service as well as economy. The greater speed, precision, and consistency of automatic control achieve more frequent service and faster rides for each passenger as well as increased total capacity for all users. If a building has too few elevators, improving their performance is a practical alternative to the cost and inconvenience of adding more cars.

Studies by the National Association of Building Owners and

Managers show average office-building elevator operating costs about half as high for operatorless as for attendant service. In apartment, hospital, and hotel buildings, which need elevator service 24 hours a day, 7 days a week, automation yields even greater operating economies.

Fig. 12.1. Elevators being modernized while one car (center) continues in operation.

Elevators are usually modernized by retaining the basic machinery and adding the necessary automatic controls. In addition to savings in cost, automatic elevator service often permits a building to earn higher rental rates at a greater percentage of occupancy. Competently planned modernization therefore represents a capital investment that can be recovered with attractive rapidity.

12.2. *Modernization Objectives*

Planning a new elevator installation is based on study of the building's height, size, design, anticipated occupancy, and other factors. Elevators must be able to handle the expected vertical traffic and provide the desired quality of service.

In addition to these factors, modernization planning must also include evaluation of the gap between present elevator performance and the capability needed to attain necessary standards. The nature and extent of required improvements therefore depend in part on the program's specific objectives.

Gaining economies from automation while maintaining present standards of service may be a primary purpose. Alternatively, a changeover may emphasize increased elevator efficiency to eliminate one or more cars. Finally, modernization may be planned to improve substantially the elevator handling capacity and service frequency.

Some buildings were underelevatored from the start. Other installations, originally adequate, fail to meet today's needs. Over the years, rising rents, coupled with improved control of indoor environment, may have led to more intensive use of space. Vertical traffic, consequently, may have grown until elevators are overloaded and service is impaired.

Traffic may have changed in pattern as well as in volume. An office building that once had a diversity of tenants, for example, may now be occupied by a few, large organizations; the result is greater demand for interfloor service than the elevators were originally engineered to provide.

12.3. *Extent of Modernization*

Without elevator attendants to adjust service to the building's requirements, performance of automated elevators depends entirely on capabilities engineered into control systems. Manually operated elevators require the most extensive modernization, including installation of selectors, controllers, door operators, and other equipment, to automate the functions of car and door operation outlined in Chapter 5.

Partially automated installations are modernized by incorporating in the control system additional equipment for fully automatic car and door operation and group supervision.

If a building has two or more elevators, further controls must be added to replace the functions of the starter in dispatching the cars of each group. Equipment for automatic group supervision needs capabilities depending on the intensity and distribution of traffic and the standard of service desired.

Elevator automation has made such rapid strides recently that marked improvement is now possible even in systems installed or modernized for "operatorless" service only a few years ago. Computer-supervised systems of group control can significantly upgrade the capability and efficiency of an elevator installation by adjusting its performance more precisely and promptly to changing traffic demands.

No two buildings, even of the same size and type, have the same elevator service requirements. A traffic survey will accurately measure existing service demands in an existing building. If comprehensive modernization may appreciably change the building's occupancy and circulation patterns, a traffic analysis, rather than a survey, will project probable service demands.

Traffic studies—surveys, analyses, or both—indicate the control system capabilities necessary for automated service in the building concerned. Comparison of these functions with those the system can presently perform guides the planning of necessary improvements.

12.4. *Improving Performance*

Manual controls may be replaced by automatic controls, or an existing automatic system may be improved to raise performance standards of the entire elevator installation (Table 12.1). Gains in service quality and quantity are most pronounced as a result of modernization programs entailing complete automation of elevators that were manually operated.

A changeover of this comprehensive nature usually includes conversion to automatic elevator control and power door operation, the latter automatically coordinated with self-leveling of the car.

Table 12.1
Functions automated in elevator modernization

Car and door operation
Automatic car operation. Each elevator is automatically controlled to respond to car or hall button calls for service in its direction of travel (Section 5.3).
Load weighing. Total passenger load in the car is weighed electrically, generating data used in the control of car operation and group supervision (Section 5.8).
Door operation and leveling. As a car stops level with the landing, power operators on the car open car and hoistway doors (Section 5.3).
Door reversal. To protect passengers, closing doors stop and reverse automatically if a person is in their path. Reversal is initiated by rubber safety shoes, light ray-electric eye arrangements, or electronic proximity detectors. To save time, doors reopen as far as necessary and begin to reclose immediately (Section 5.4).
Nudging. Prevents service delay by a passenger who blocks doors and causes reversal devices to operate. After a predetermined interval a warning buzzer sounds in the car, doors close slowly but safely to nudge obstruction aside, and signal notifies building personnel of potential delay (Section 5.4).
Group supervision
Service for moderate traffic. Two or more elevators are automatically coordinated by multiple zoning so that each answers its own car calls and hall calls from certain floors. Floors are separated into zones and cars are distributed among the zones for prompt response to calls (Section 5.7).
Service for intensive traffic. Elevators operate continuously at intervals automatically controlled to serve all passengers in minimum time (Section 5.8).

Because the control system now starts and stops the elevator and closes and opens its door with greater precision and promptness, a typical stop-to-stop time may be reduced from 20 sec. to 14 or less.

Along with automation of car and door operation, a group of two or more elevators requires installation or modification of an automatic supervisory system to attain required performance capabilities. In some buildings—apartment houses and suburban motels and smaller hospitals and office buildings—control and supervisory systems need to be capable of handling only moderate traffic (Section 5.7).

In many office buildings, hospitals, and hotels, "heavy" elevator traffic, when many users must share available elevator capacity, characterizes part of the working day. Automating elevator operation in these buildings calls for group supervisory systems (Section 5.8) which economically handle heavy traffic by concentrating service where demand is greatest.

By quickening the pace of elevator operation and coordinating the distribution of elevators in a group so that cars are closer to calls, modernization can appreciably reduce waiting and riding time.

Table 12.2
Handling capacity increased by faster round trips

The number of people (C) an elevator can handle during each 5-minute period equals the time, in seconds, multiplied by the number of people the car will accommodate, divided by round-trip time, also in seconds. Reducing round-trip time from 120 to 100 sec. increases 5-minute handling capacity from 40 to 48 passengers, or by 20%:

$$C = \frac{300 \times 16}{120} = 40$$

$$C = \frac{300 \times 16}{100} = 48$$

Service is not only improved for each passenger but, because elevator round-trip time becomes shorter, the cars can also carry a greater number of passengers in a period of time (Table 12.2). Improving performance by 20 percent, as in the example in Table 12.2, is like adding one extra elevator for every five already installed.

12.5. *Modernizing Cars and Entrances*

Modernization is usually planned to update elevators not only in performance but also in appearance, comfort, and convenience.

"Visual modernization" includes new elevator cars and entrances incorporating features reviewed in Chapter 11, designed to harmonize with redecorated building lobbies and corridors (Fig. 12.2).

When new cars are installed in existing car frames, rubber isola-

Fig. 12.2. Contrast between elevators in foreground and still-to-be-modernized cars in background dramatizes updated appearance as well as performance. New elevators cars and entrances accompany changeover of control systems for fully automatic operation.

tion cushions noise and vibration. Rubber-tired roller guides may replace sliding-type shoes to improve riding quality.

Close-coupled cars and entrances save space to permit platforms of increased capacity in existing hoistways. If hoistway conditions permit, center-opening entrances may be substituted for single-slide or two-speed doors. New hall lanterns and in-car signals guide passengers to elevators going their way. As a result of improvements like these, people enter and leave cars faster, speeding the general pace of service.

Conversion of elevators to unattended operation often includes installation of an intercom system with a loudspeaker and sensitive microphone in each car to keep passengers in touch with the building staff.

Properly applied automation may increase elevator plant efficiency in an existing building and permit elimination of one elevator in every group. Recaptured hoistway space may then be used to improve corridor layout or to accommodate air-conditioning ducts or electric power risers. If, on the other hand, new air-conditioning and lighting will result in more intensive use of space, total elevator handling capacity should be increased accordingly.

Fig. 12.3. In use for forty years, escalators in a large New York store were still serviceable but outmoded in appearance and safety.

Fig. 12.4. Retaining the original truss (diagonally across the photo, from upper left to lower right) permits modernizing escalators in minimum time and with minimum building work. Exterior finish, wellway railing, and fire-barrier kiosk are not disturbed during the changeover.

12.6. *Escalator Modernization*

Modernizing elevators in an existing building amounts essentially to installing new control systems and possibly new cars and entrances. But comparable improvements in escalator service and appearance formerly required substituting completely new escalators for old. Crews had to remove and replace each escalator truss, a structure some 50 ft. long and weighing approximately 4 tons.

Aside from demolition and construction, removal and replacement of escalator trusses created severe logistical problems in an occupied building. Old and new structural elements had to be moved through the building and stored, loaded, and unloaded in locations where space is at a premium.

Procedures have recently been developed that leave the truss in place but replace all parts that affect appearance or performance. Escalators are stripped down to the steel members of the truss, which is then re-equipped with new running gear, steps, and balustrades (Figs. 12.3, 12.4, and 12.5).

Retaining rather than replacing trusses permits escalator moder-

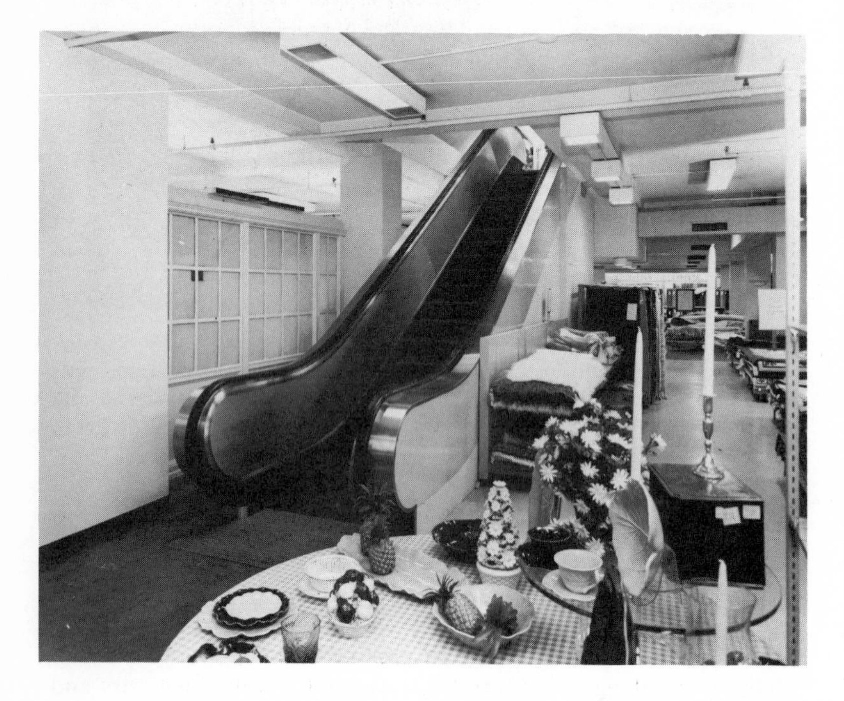

Fig. 12.5. Escalator modernization incorporates current practice in design, construction, and materials. Balustrades are attractive, easily cleaned stainless steel, newels are extended to ease passenger entry and exit, hand rail brushes are recessed away from children's finger tips.

nization with minimum down time and minimum inconvenience to a building's occupants and visitors during the changeover period.

These methods save costly, extensive building work. Along with the truss, adjoining structural elements, such as soffits and soffit lighting, side coverings, sprinklers, and electrical conduits, are left in place. Existing electrical feeders are simply reconnected to the controller of the new escalator machine.

By eliminating major construction, the simplified method significantly shortens the time schedule for escalator modernization. The work not only goes faster but, while under way, interferes less with normal building operations.

Planning elevator and escalator modernization in coordination with other improvements in building facilities and decor can hold down total costs and enhance the effectiveness of the entire program. Modernizing vertical transportation in an existing building presents the architect or engineer with a challenge equaling or exceeding that of designing an entirely new system for a building still on the drawing board.

Chapter 13

Planning for Vertical Transportation

In terms of building design, economics, and operation, vertical transportation serves the basic purpose of making space in a multistory building as accessible as possible, by taking people to their floors safely, promptly, and comfortably, with long-run economy. While many installations fulfill these objectives to the satisfaction of architects, owners, and users, others fall far short of optimum goals.

Once a building has been completed, elevators or escalators can hardly be added unless advance provision was made for extra hoistways or wellways. To avoid costly mistakes that may persist for the lifetime of the building, vertical transportation demands competent consideration early in planning a construction or modernization program.

Preceding chapters have outlined the equipment that constitutes an elevator or escalator system, its place as an integral element of a multistory building, and its function in performing an essential service for the building's users. Understanding these fundamentals, the architect or engineer can better evaluate plans of an actual system for a proposed building or the improvement of an existing installation.

13.1. *Evolution of Planning Methods*

Effective planning of a vertical transportation system depends on ability to analyze a job's specific service requirements as well as on competence in using the means available for satisfying those requirements. With increasingly complex requirements as well as means, planning procedures have evolved from rough empirical methods to highly precise analytical techniques made practical by progress in electronic data processing.

Increasing use of digital computer procedures in elevator and escalator system design characterizes the latest of three periods in the development of vertical transportation planning.

The first historic period, one of pioneering, began with the safety elevator's original appearance in mid-nineteenth century and continued through the first quarter of the twentieth. During most of this time, merely carrying people up in a building was deemed an achievement. Gradually, increasing experience led to rule-of-thumb methods for designing elevator plants approximately sized to service needs.

By the 1920s, engineers had accumulated sufficient experience to formulate general criteria for analyzing traffic demands in a projected building and calculating the performance of its elevator or escalator plant. They needed more dependable system design procedures to satisfy the more critical requirements of the period's taller buildings and properly to apply the higher-speed elevators with partially automated controls then becoming available.

These circumstances led to the development of simple equations expressing relationships between traffic demand and system capability in terms of service quality and quantity (Section 3.8). Using observed, derived, or assumed values, traffic engineers could calculate anticipated elevator service and predict with adequate accuracy how well vertical transportation quantity and quality would satisfy the needs of a building and its occupants.

13.2. *Analyzing Elevator Working Cycles*

Engineers usually estimate elevator round-trip time by adding together the time periods required for each of the consecutive operations in a complete trip. Once average round-trip time is known, carrying capacity and average waiting time are simple to calculate.

To understand readily the essentials of calculating round-trip time, consider the simplest case, an automatic elevator serving only two stops and carrying only one passenger from the lower to the upper stop (Table 13.1). At the start of the working cycle, assume that the car is parked with its doors closed at the lower landing.

From the moment the passenger touches the hall button at the lower landing until he leaves the elevator at the upper floor, a total of some 19·5 seconds (total "up" trip time) usually elapses. It takes another 10·5 seconds to return the car to the lower landing in readiness to serve similarly a second passenger, bringing "total round-trip time" to 30 seconds.

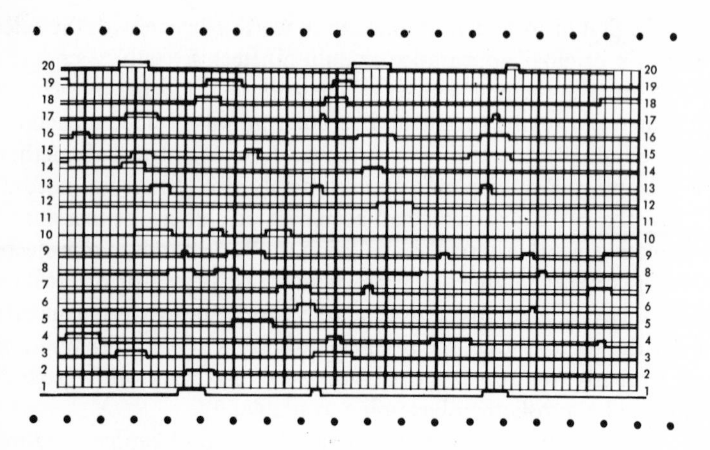

Fig. 13.1. Graphic analysis of elevator traffic and performance. An event-recording meter plots time for which calls from up and down hall buttons at representative floors remain registered. Fine vertical lines divide chart into 10-second periods, and trace indicate duration of calls by rising above base lines. This chart covers 12 minutes of moderate traffic in a downtown hotel.

Table 13.1
Calculating round-trip time for two-stop elevator

Passenger touches hall button, door open	3 sec.
Passenger enters car, touches car button	2
Doors close	3
Car travels to upper landing	7·5
Car levels and doors open	2
Passenger leaves car	2
Total "up" trip time	19·5 sec.
Doors close again	3
Car returns to lower landing	7·5
Total round-trip time	30·0 sec.

In the example, allowances for door closing, car travel, and the other elements of round-trip time are approximations based on typical service conditions. In actual practice, time requirements depend on performance capabilities of the elevator installation and variations in the demands on those capabilities.

For instance, if the elevator carries not one but several passengers per trip, more time must be allowed for them to enter and leave the car, thus increasing one-way round-trip times. By observation and calculation, engineers have determined how many seconds to allow for passenger transfer under various loading conditions.

Calculations also become more complex when an elevator serves, not only two stops, but several, as is more often the case. On any one trip the elevator probably stops only at some but not at all of the floors it can serve. Probable stops depend on such factors as the number of passengers in an elevator and the distribution of population among the floors of the building.

Calculations are also influenced by the type of traffic under consideration. Other variables include the number of occupants in a building and visitors to it and their habits of floor-to-floor movement and elevator usage.

In an existing building where elevators are to be modernized, much of the basic traffic data can be determined by observation. But, in a proposed building, data must be projected from experience. During recent decades, elevator engineers have accumulated fairly dependable traffic and performance data for a wide range of conditions and have developed methods of handling the data.

13.3. *Planning for Service*

Further experience confirmed the essential validity of these principles of traffic analysis and system design and permitted continuous improvement in their application. Beginning in the 1950s, however, rapidly changing conditions called increasingly for planning procedures of greater precision and dependability. As data processing technology advanced, it promised to provide powerful new tools with which vertical transportation engineers could satisfy even more stringent demands.

In office buildings, traditional methods of calculating traffic during critical peak periods are relatively simple and usually yield reliable results for determining the carrying capacity of an installation. But these methods may not provide service of adequate quality for patterns of interfloor traffic that often prevail in office buildings as well as in other types. Adequate analysis of these more complex forms of traffic by conventional, manual methods would require so many observations and calculations as to be prohibitively time-consuming and costly.

Trends in building design and use (Chapter 10) are making patterns of vertical circulation more complex in the distribution of traffic demand from floor to floor and its fluctuation from moment to moment.

Towers are assuming new shapes and soaring to new heights. Apartment, office, and retail store levels may be stacked vertically in the same building. Upper- and lower-floor traffic generators, like rooftop cocktail lounges and basement parking garages, impose concentrated, fluctuating loads on vertical transportation. People come and go more frequently, while large organizations occupying several floors of a building create concentrated demand for localized service.

13.4. *Planning for Economy*

With vertical traffic heavier in volume and more complex in distribution, cost-conscious owners insist on faster, more frequent floor-to-floor service for themselves or their tenants.

In a typical office building, each elevator may serve 250 to 300 people. Each person uses the elevators about six times a day and takes an average of 50 seconds per trip for waiting and riding time. On the basis of the lower figure, 250 persons per elevator, time spent waiting for and riding on elevators would total 5,000 man-hours a year for each elevator in the building (Table 13.2).

Improved service may save 20 percent of the total, or 1,000 man-hours. The time saved may be evaluated on the basis of the full hourly cost of personnel, from errand boys to executives, to the organization employing them.

Since an elevator or escalator system to satisfy today's service demands may represent some 10 percent of total construction cost for an office building, owners understandably expect maximum return from their vertical transportation investment. More efficient systems release usable space and contribute to greater economy in the design of the building as a whole.

Table 13.2
Time consumption for elevator trips

Each person takes	6	elevator trips a day
Each trip consumes	50	seconds waiting and riding =
Each person spends	300	seconds a day on elevator trips ×
	250	people using each elevator =
Time consumed per day per elevator	75,000	man-seconds or more than
	20	man-hours ×
	250	working days a year =
	5,000	man-hours per elevator per year

With complete automation of elevator control, system designers could no longer rely on attendants and starters to adjust service to possibly unforeseen traffic conditions. Since planners now had to anticipate potential situations and engineer appropriate capabilities into the control system, precision in traffic analysis and system design became more urgent.

With increasingly critical standards of traffic analysis and system design, types of equipment that could be combined in varying ways for an optimum solution grew more numerous.

Besides deciding on the number, size, and speed of units, engineers also had to determine if a building might be best served entirely by elevators, by escalators, or by a system combining both forms of transportation. Various control and supervisory systems were now available to operate elevators and groups in response to service demands.

Choosing the most economical system would require comparing performance and cost for many possible alternatives. The length, complexity, and expense of the necessary calculations tempted designers to oversimplify problems by excluding all but a few, seemingly most influential factors; the result was often a less-than-optimal system.

Essentially, quality and quantity of elevator service in a building depend on: the traffic, people demanding transportation; and the elevators, their operation in response to passenger demands. To gather all relevant data on traffic demand and system performance and analyze them for system design, more refined procedures had to be developed.

13.5. *Data Collection and Analysis*

Most traffic data were formerly recorded by trained observers with stopwatches and notebooks. Data collected in this way may be processed electronically, on digital computers, or used in manual calculations.

But input data from visual observation are limited at best, since human observers can follow only a few variables, can record values only at intervals, and are subject to fatigue and other failings. Visual observation, therefore, is being supplemented by recording instruments that provide more accurate, continuous, and comprehensive data on passenger demands and elevator performance.

Special-purpose instruments like people-seconds meters help observers to measure waiting and riding time for passengers. Event-recording meters automatically chart elapsed time from registration of a call until it is answered, and duration of elevator runs from stop to stop and time spent at each stop. Graphic analysis is one means of using the more comprehensive information provided by these instruments to study the people-elevator interaction in direction and distribution over a period of time.

For such an analysis, elapsed-time or event-recording meters may be connected to record actual periods during which calls at various floors remain registered. From one of these hall-call registration charts (Fig. 13.1), engineers can calculate average waiting time during any period. The chart also shows the floor-by-floor distribution of service demand at any instant.

13.6. *Computer Simulation*

Design of an optimum system requires more comprehensive analysis of the relationship between anticipated traffic and elevator system performance, accounting for all relevant variables. This can be accomplished by simulating the people-elevator interaction with realistic models, each an interrelated logical, mathematical representation of a complete vertical transportation system in its building. Rapid, accurate calculation and comparison of alternative models become practical with high-speed digital computers.

For the "people" element of the model, data from traffic-flow patterns in buildings of the type planned are used to program the computer (Fig. 13.2). Accompanying input includes parameters reflecting the design of the particular building and its expected

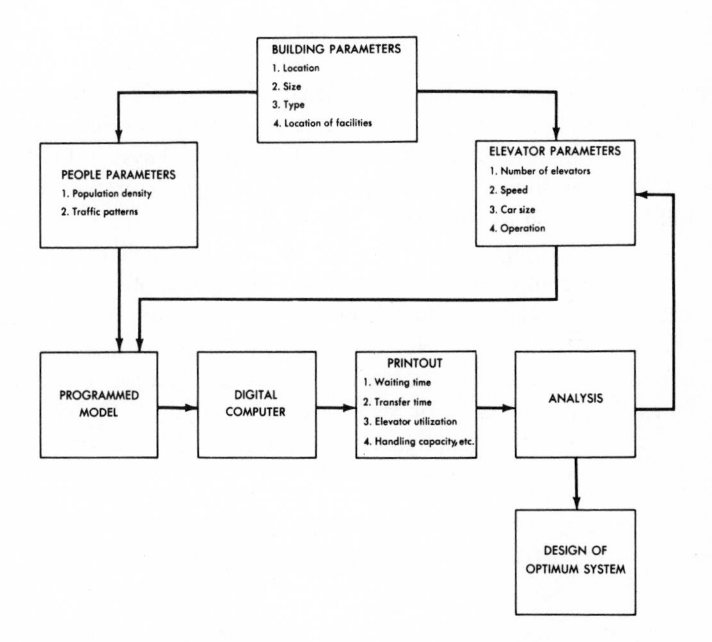

Fig. 13.2. Steps in designing a vertical transportation system with a logical, mathematical model programmed on a digital computer. The model realistically simulates, for a variety of different conditions, the interaction between people and the elevators or escalators they use.

occupancy. Service demand can be simulated for different conditions of weather, season, horizontal transportation, and other elements.

To simulate the "equipment" factor the computer is also fed data on the number, size, speed, and other pertinent parameters of the elevators or escalators. Computations are run for various building designs and methods of automatic elevator operation and group supervision (Fig. 13.3).

For each set of simulated conditions the computer prints out data analyzing elevator performance in far greater significant detail than was previously possible. Equipment and construction costs can also be computed for various alternatives.

Results are more realistic as well as more complete. Many determining variables can now be readily evaluated, and the model can continuously simulate the dynamic effects of elevators and passengers on one another. The speed of electronic computation readily permits projecting detailed data on elevator service for each of many combinations of elevator plant and control system factors.

Computer simulation permits making decisions on elevatoring a building early in the planning stage with confidence that service in the completed, occupied structure will be close to calculated values.

Study of computer-simulated models proved helpful in planning elevator and escalator installations for the 110-story World Trade Center tower buildings in New York, which set precedents in height and design and therefore made it unreliable to project experience from earlier skyscrapers. For the future, this analytical approach also promises to reveal new design and performance criteria for vertical transportation systems.

13.7. *Scheduling Elevator Installation*

Architect responsibility for vertical transportation extends beyond integration of an elevator or escalator system with building design to scheduling and supervision of equipment installation as a critical element in high-rise building construction.

Elevator construction must be scheduled to mesh smoothly with progress by other trades so that neither delays the other. As the general contractor builds each hoistway, the elevator contractor installs buffers in the pit, guide rails along the walls, and overhead machinery and equipment in a machine room at the top.

So that each piece of elevator or escalator equipment is ready for installation at the scheduled time, manufacturing has to be started months earlier and engineering even further in advance. Deliveries must be made exactly on schedule, because downtown sites leave little room to store equipment arriving ahead of time, while delayed delivery is costly. In a tight labor market, elevator constructors and other specialized skills must be rounded up as needed.

Bar charts have long aided architects and engineers in planning these schedules, but the Critical Path Method, a computer-age

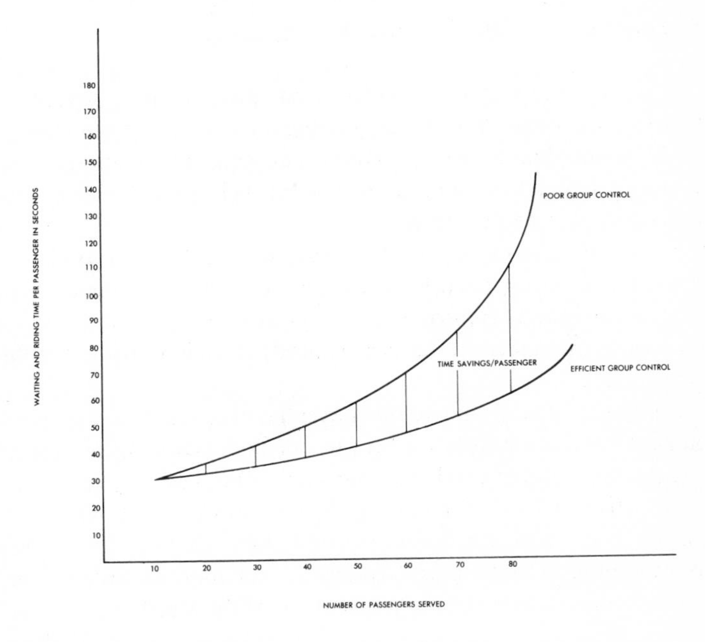

Fig. 13.3. A wealth of pertinent cost and performance data may be derived from computer simulation of elevator-people reactions in possible vertical transportation systems. Here two group supervisory systems are compared in terms of person-seconds spent by all passengers waiting for and riding on elevators in a building.

development, is proving a more useful tool for the purpose. The CPM network diagram graphically integrates all key activities so that they may be more readily brought under control to anticipate and eliminate possible delays.

Elevator work is correlated with other construction on the

diagram. Schedules may be calculated manually or, for a more complex project, on an electronic computer. From the CPM network and schedules the architect, engineer, and contractor gain a comprehensive view of significant interrelated factors on which to base decisions with confidence.

13.8. *Toward More Effective High-Rise Buildings*

New, more powerful tools of analysis promise more perceptive insight into the relation of building design and occupancy to circulation and traffic. Planning of buildings that gain full advantage from high-rise design, yet minimize demands on their vertical transportation systems, will be facilitated.

Future developments in elevators and escalators promise to follow directions suggested by accurately analyzed requirements for service and economy. Improved equipment will be applied in systems more precisely designed to the specific needs of each individual building and its users.

Computer simulation and other avenues of research and development seek continuing improvement in our understanding of vertical transportation fundamentals incorporated in improved systems, and more effective methods for putting them to work. Those who plan, construct, own, and use multistory buildings can look forward to their becoming at once more serviceable yet more economic, and therefore more valuable in the truest sense of the word.

Index